Grade 4
Mathematics
Teacher's Resource Book

PAT LILBURN
PAM RAWSON
PETER SULLIVAN

in consultation with
Elsie Kinavai
Alex Feeger

Department of Education Papua New Guinea

OXFORD
UNIVERSITY PRESS

OXFORD
UNIVERSITY PRESS
AUSTRALIA & NEW ZEALAND

253 Normanby Road, South Melbourne, Victoria 3205, Australia

Oxford University Press is a department of the University of Oxford. It furthers the University's objective of excellence in research, scholarship, and education by publishing worldwide in

Oxford New York

Auckland Cape Town Dar es Salaam Hong Kong Karachi
Kuala Lumpur Madrid Melbourne Mexico City Nairobi
New Delhi Shanghai Taipei Toronto

With offices in

Argentina Austria Brazil Chile Czech Republic France Greece
Guatemala Hungary Italy Japan Poland Portugal Singapore
South Korea Switzerland Thailand Turkey Ukraine Vietnam

OXFORD is a trademark of Oxford University Press
in the UK and in certain other countries

First published 2000
Reprinted 2008

ISBN 978 0 19 550864 2
ISBN 0 19 550864 5

Written by Pat Lilburn, Pam Rawson and Peter Sullivan
Writing consultants: Elsie Kinavai and Alex Feeger of the Curriculum Development Division of the Papua New Guinea Department of Education
Editorial and design consultant: John Hughes of the Papua New Guinea Department of Education

Cover and text design by Mary Kerr
Cover artwork by Gigs Wena
Illustrated by Wendy Gorton
Printed in China by Golden Cup Printing Co Ltd
Published by Oxford University Press
Editorial Office: PO Box 7979, Boroko NCD,
Papua New Guinea

CONTENTS

SECRETARY'S MESSAGE

This teacher's book is part of the new reformed mathematics program designed and written for use in Grade 4 classes in Community and Primary schools throughout Papua New Guinea.

This teacher's book is accompanied by two pupil books called *Grade 4 Mathematics Pupil Book* 4A and *Grade 4 Mathematics Pupil Book* 4B. They replace the MACS series now in use.

In the Mathematics program children are first taught mathematics by using real objects. Later, the children will use pictures of objects. Finally, the children will use number symbols to represent these objects. Always allow your pupils to use real objects during their mathematics lessons if they want to. Teach the concepts in the context of real life situations as this leads to an appreciation of the everyday use of mathematics skills and knowledge. The children will decide when they do not need the help of real objects any longer. Remember, when children use real objects, they will understand mathematics better. This program also encourages the children to solve their own problems and make their own decisions with confidence. In order to learn these skills, the children should talk about what they are doing in every lesson. Since learning is most effective when it has meaning and is enjoyable, allow the children to use the language that they are most comfortable with.

However, in Grade 4 more English should be used together with the vernacular and other languages to aid understanding of new and difficult concepts.

Finally, the National Department of Education wants teachers to be flexible in programming and timetabling to cater for the different abilities of children and multigrade teaching situations. This book shows you some strategies to help satisfy these needs.

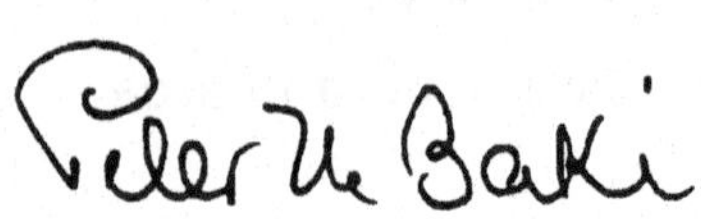

PETER M. BAKI
Secretary of Education

INTRODUCTION

A/W 1a (TEACHER'S RESOURCE BOOK 4, PUPIL BOOK 4A, PUPIL BOOK 4B)

HOW TO USE THE TEACHER'S RESOURCE BOOK

THE UNITS

The Grade 4 Program is divided into 34 units. These units are the basis of this Teacher's Book, and follow the scope and sequence in the syllabus. Units 1 to 18 can be taught in terms 1 and 2, and Units 19 to 34 can be taught in terms 3 and 4. The units can be taught in order, because the lessons have been organised this way to make learning as easy and effective as possible. However, you may change the order to suit your children's needs. Basically, each unit consists of a set of objectives and a collection of relevant activities that can be used to teach the objectives.

Unit 34 has been written in a different format to the other units. This unit uses the theme of *Gardens* to bring together much of the mathematics that has been taught during the year. It attempts to place this mathematics into a practical and familiar setting so that children have the opportunity to explore and use mathematical concepts in real situations. Investigation and problem solving are emphasised in this unit. You should allow the children plenty of time to explore the mathematics of each situation.

Information at the start of each unit

At the start of each unit you will find a page of information which briefly explains what the unit is about. The **title** indicates the part of the curriculum which the unit covers. There is a brief summary of what **the unit builds on**. This may refer to the relevant section of *Resource Book* 4 or to some earlier unit. A list of the key new **English words and expressions** used in the unit are given, in the order that they appear. There is a brief summary of the **main ideas** in the unit to show the detail of what is being covered. The **objectives** list the experiences and outcomes for the pupils. In many units some **special methods and strategies** that can be used in the unit are listed.

Activities within each unit

Within each unit there are a number of activities for teaching each objective. You can select the activities from the ones provided which best suit your class, and what they have done already. There are too many activities to do in the time available, and so you will need to choose. Some of the activities are essential, so you should aim to do all of these. You can then make a selection from the remaining activities, based on the time you have left, and the needs of your class. Some activities refer you to the Pupil Books. Many of the activities will take only part of a maths lesson, so sometimes you may need to plan to use a number of activities in any one lesson. You are also encouraged to adapt any of the activities to best suit the needs of your pupils. There are also some games in the Grade 4 activities. Games are an enjoyable way for children to learn and practise mathematical concepts and skills. The first time a game is mentioned in a unit, the rules are given. After that, you are referred to that first mention.

The different activities within each unit have been identified as follows.

- activities which are boxed are essential activities which should be used
- activities which are shaded yellow are pupil activities requiring minimal teacher input.

The following diagram gives an explanation of the coding system used in the book.

SAMPLE ACTIVITIES

OBJECTIVE 1.1

Annotation	Activity	Description	Materials	Annotation
name of activity; new word (in bold) to teach children	**Order of Digits** (new word)	Write 82 on the board and ask a child to show it using shells, sticks or base ten blocks. Write 28 on the board and ask a different child to model it. Ask how 82 is different from 28. Even though the two numbers both have an 8 and a 2, the order of the **digits** makes them different and the right hand **digit** tells us how many ones in the number and the other **digit** tells us how many tens it has.	bundled sticks, base ten blocks, strings of shells or any material that can be used for modelling numbers to 100	
teacher activity	**Four digits**	Write four numbers from 1 to 9 on the board, e.g. 5, 9, 2, 1. On a piece of paper, write a number made up of these four digits, e.g. 1952. The children guess what number you made and write it on a piece of paper. Show the children your number and ask them to check their number. Discuss how four digits can make many different numbers. Let the children do this activity in groups.	small pieces of paper	
pupil activity	**Writing numbers**	Make two teams. Have two tables with small pieces of paper and a pencil on each one. Write a different four digit number on two separate pieces of paper and place one on each table, face down. The first member of each team comes to their table, reads the number, copies it onto a new piece of paper and places this face down, hands the original to you and returns to their place. Repeat until all children have had a turn. Compare the final number written by each team with the original number.	small pieces of paper, two tables and pencils	materials needed for activity
essential teacher activity	**Number strings**	Ask four or five children to come to the board. Read out a string of four numbers, e.g. 2, 5, 9, 7 and let the children write them. Compare the numbers that the children wrote. Were they the same? Repeat 3 or 4 times. Repeat the activity using strings of 5 numbers then strings of 6 numbers.		
essential pupil activity	**Matching numbers**	The children can answer the questions on the page. Make up some more questions for the children to answer.	4A, p 1	Pupil book 4A page 1

PLANNING YOUR TEACHING

There are 210 minutes per week allocated for teaching mathematics. You may plan to teach 7 thirty minute mathematics lessons each week. It is not necessary to use the activities from the book in every lesson. A general guide is to use 5 or 6 lessons each week for the activities from the books. You can use the remaining time allocated per week to help your pupils complete, revise or extend the activities that you have already taught. In particular, you will need to decide what extra practice to give your children in the times tables and ensure that they can confidently use the basic operations. You must decide how you would like to do this.

You may choose to spend about one week on each unit. In this way you may have some weeks at the end of the year for revision as well. Since it is now the National Department of Education's policy to be flexible in timetabling and programming, the books have been written to help you to do this. Remember to use this to your advantage!

HOW TO USE THE PUPIL BOOKS

There are two Pupil Books for Grade 4. **Book 4A** is for terms 1 and 2. **Book 4B** is for terms 3 and 4. The books are used to support the mathematics program and should not be used by themselves. You should have enough Pupil Books to share one between two children. These books are **not** to be written on. Usually, the activities that refer to the books ask the pupils to copy something into their workbooks and to complete the exercise or to solve a problem. You may need to help some of your children understand some of the pages. Some pages in the Pupil Books are used in only one activity, others are used in several different activities.

HOW TO CARE FOR THE BOOKS

You must be careful to store these books in a dry, safe place. Rain, ants, cockroaches, silverfish, mice, rats and other animals will destroy the books if you and the children do not take care of them. Use a clean, dry box with a lid to store the books in. Place a few moth balls in the box to kill the insects. You might also cover each book with plastic or newspaper to keep them from getting dirty or ripped. Dust them often if they are stored for more than a month at a time.

LANGUAGE

Grade 4 is the first year that most teachers will use a lot of English as the language of instruction, with vernacular continuing to be used to introduce new concepts and methods, facilitate understanding and reinforce meaning. Encourage the use of both languages. Each unit contains key words that need to be introduced in English. These new words are marked in bold within the activity and signposted in the margin. The first few times children meet these new words you should:

- Say the word with the class a number of times
- Write the word on the board
- Explain the meaning of the word using real objects, actions, pictures
- Demonstrate how to use the word in a simple sentence
- Ask the children to use the word in a simple sentence
- Tell the children to enter the word into their *Language Vocabulary Book* or *Class Dictionary*.

MATERIALS

The Grade 4 Mathematics Program uses many different materials. It has been shown that children learn to understand new concepts when they use real materials at the same time as they are introduced to the new idea. Many of the materials that you need for the mathematics program are things that can be collected by you and the children. Before children use new materials they need time to explore them. This will help them concentrate on the tasks that you set instead of wanting to play with the new materials. Some time is allowed for this in the lessons where new material is introduced, but you may wish to allow more time. Children should be encouraged to use the materials whenever they need them. Children must know where to find and return materials. Storing the materials on shelves, in strong boxes which have lids, makes it easy for children to use them. Make sure that the children know that you expect them to look after the materials. Check all materials regularly and repair or replace any that are damaged. Some of the materials that you should have available in your classroom are listed below.

Collections of objects

In many lessons the children need objects for counting, matching and measuring. These can be small stones, buttons, shells, bottletops, seeds or similar items. Store them in suitable containers.

Small number cards

You will need many small cards for writing numbers on. You will find instructions for making these on page 215. When you write numbers on cards, labels, the blackboard, or anywhere else, make sure you take care to write them properly. Refer to page 220 which shows you the correct style. Children copy your example. If you are careless, they will be too. Always start the numbers in the correct place and take care with their formation.

Time table cards

You will need a set of cards for all the tables. These should be no smaller than 3 cm X 8 cm. You will find instructions for making these on page 215.

Dice

Make a dice or a hexagonal spinner, for each child from wood or card. You will find instructions for this on page 215.

Balances

These are needed in Unit 13. If they are not available in your school you will need to make some. You will find some plans for simple balances on page 216.

Containers

You will need to collect a variety of these, especially for the capacity activities in Units 15 and 29, and the fraction and weight activities in Unit 4. Read these units carefully so you know what you need to collect.

Base ten blocks

Each set contains four different kinds of wooden blocks. The picture below shows the different blocks. The smallest one is called **a cube**. The next one is called **a long**, the next is called **a flat** and the largest one is called **a block**. At least one set has been supplied to each school.

Place value squares

Page 217 of this book shows you how to make these. You will need to use them often, so make them with firm paper or card and store them carefully.

Bundles of sticks

Use split bamboo or broomsticks about 10 to 15 cm long to make bundles of ten sticks. Also have many single sticks not in bundles.

Spear cards You will need several sets of these. They are used many times throughout the year. You will find instructions for making them on page 217.

Demonstration clock Make a demonstration clock. You will need this in many units. You will find instructions for making this on page 217.

Coin rubbings Make many rubbings of each of the coins. There are instructions for how to do this on page 216. (You may want the children to help you.) Plastic PNG play money coins have been supplied to grades 1 and 2 and can also be bought from school suppliers.

Tangrams You will need 5-piece tangrams. You will find instructions for making these on page 217.

Angle testers Each child will need to make an angle tester in Unit 9. Instructions for making these are on page 218.

Nets In Unit 16 you will need a triangular prism and a hexagonal prism. There are nets for you to copy and instructions to make these on pages 218-219.

ASSESSMENT

Assessment is the process of finding out what children know and what they have learnt. It is important that you know what mathematical concepts the children understand, how they learn best, and what concepts they are ready to learn. It is not necessary (or appropriate) to use only tests to find out what the children know. Good assessment information can be gathered during everyday teaching by listening to what the children say, by watching what the children do, and by looking closely at what the children make, write and draw.

WHEN AND HOW TO EVALUATE YOUR PUPILS

You should continually build up your knowledge of what the children know. This starts on the first day at school and continues through the year. The ongoing assessment of learning is a natural process which all teachers can do. Some important ways for you to learn about what the children can do are:

- by listening to the children's answers when they respond to questions
- by watching carefully what the children do during classroom tasks
- by listening to what the children say when they are talking to each other and to you

It is important that you do not expect children to have mastered all the concepts at this stage. If you provide them with the necessary experiences they will begin to use the concepts more often.

There are some teaching strategies which will help you to get meaningful information about the children. One important way is to give *open*, rather than *closed* tasks. For example:

$\square + \square = 43$

is an *open* task. There are many possible correct answers. The children can find some different number combinations which can be added to make 43. In so doing, they may become aware of the pattern involved when writing answers, e.g. 1 + 42, 2 + 41, 3 + 40, etc. You can learn about their understanding from their response.

On the other hand:

$25 + 18 = \square$

is a *closed* task. The children only need to perform the addition process to find the answer. Such questions are useful at times, but teachers do not learn much about the children from such questions.

It is also important that you give children the opportunity to talk about mathematics to each other and to the class. This is a good way to learn, since language is one of the first steps in forming mathematical concepts. Listening to children also tells you much about what the children know. Another way to find out more about your pupils is to look closely at their drawings and written work.

Try to collect some information during each mathematics lesson. It is better to collect a small amount of information each time rather than trying to collect a lot all at once. One way to do this is to select three or four children in a particular lesson and observe them. Listen to their answers and look carefully at their work.

HOW TO RECORD ASSESSMENT INFORMATION

Teachers who watch and listen to children soon learn a lot about what the children know. It helps if there is a systematic way for recording this information about the children. You should choose at least one method and stick to it. You may decide to use a combination of two methods. The most important part is to keep your records up to date. There should be some change almost every week.

Samples of student work

Some teachers use a collection of each child's work as a record of that child's performance. You keep a folder of work for each child. Every time a child produces a new piece of work they discuss with you whether or not it should go into their file. Only pieces of work which show new understanding or improvement on old skills belong in the file. If it is an improvement on older work, then the older work is removed from the file to stop it getting too full. This only works well if:

- the children are actively involved in choosing what is in the file
- the files are kept up to date
- enough suitable student work is produced. (If you are short of paper, much of the children's work will be on blackboards which cannot be put on file.)

Checklists Make a check list of the main objectives from the scope and sequence. You should make entries in the checklists progressively. As each bit of information is found out, a note should be made on the sheet. It is best that a consistent method is used for recording the information.

A system such as the following is helpful. Use a / if a child shows you that he or she understands the concept. Use a X if the child can use the concept to do something else or solve a problem. If you use this method you must keep it up to date.

Annotated class lists Another strategy is for teachers to have a blank class list on which they can record information. It is best if this class list is used only for things which the teachers see which they do not expect. For example, a child who usually does not answer questions might give a good response. This would be worth recording on the class list.

The knowledge about the children in the teacher's mind is then combined with the information on the class list to produce a better picture of what the children have learned. It is very important to be consistent and record all unusual events. An example of an annotated class list is as follows:

Name	Comments (unexpected events)
Abat	Worked well in his group. Gave the group report.
Julie	
Ben	Did not know his 3x table.
Carrie	Understands fractions well.

USING ASSESSMENTS

Teachers assess children so that they can help the children to learn better, especially in the early grades. All assessment information should be used for teaching purposes. In other words, if a teacher finds through assessment that a child or a group of children cannot do something, then they should plan to reteach this particular idea.

If a child does not understand a concept then it is necessary to try a different explanation. It is not worth collecting assessment information if it is not used. It is only when a child understands a particular concept that he or she can be asked more difficult questions.

UNIT 1

NUMBERS IN THE WORLD

This topic builds on all number work from the previous year including ordering numbers, number patterns, and writing numbers in words, (see *Teacher's Resource Book* 3, Unit 3).

ENGLISH WORDS/EXPRESSIONS USED

digit, increased, estimate, more than, less than, strategy, predict, bigger

MAIN IDEAS

- **place value—position of digits determines value**
- **numbers can be used as counting numbers and as labels**
- **number sequence—numbers have a natural order**
- **numbers can be given in words and symbols**
- **estimation—a quick informed guess of the result can be used as a validity check on the final result**

OBJECTIVES

The children will:

1.1 compare numbers and see how they are different from each other

1.2 recognise numbers in the world around them

1.3 order numbers to 999

1.4 practise reading and writing numbers in words

1.5 estimate and count large numbers of objects

SPECIAL METHODS AND STRATEGIES

The key to this unit is for the pupils to see numbers as real and as part of their experience. Use real life examples wherever possible.

OBJECTIVE 1.1

Order of Digits

new word

Write 82 on the board and ask a child to show it using shells, sticks or base ten blocks. Write 28 on the board and ask a different child to model it. Ask how 82 is different from 28. Even though the two numbers both have an 8 and a 2, the order of the **digits** makes them different and the right hand **digit** tells us how many ones in the number and the other **digit** tells us how many tens it has.

bundled sticks, base ten blocks, strings of shells or any material that can be used for modelling numbers to 100

Four digits

Write four numbers from 1 to 9 on the board, e.g. 5, 9, 2, 1. On a piece of paper, write a number made up of these four digits, e.g. 1952. The children guess what number you made and write it on a piece of paper. Show the children your number and ask them to check their number. Discuss how four digits can make many different numbers. Let the children do this activity in groups.

small pieces of paper

Writing numbers

Make two teams. Have two tables with small pieces of paper and a pencil on each one. Write a different four digit number on two separate pieces of paper and place one on each table, face down. The first member of each team comes to their table, reads the number, copies it onto a new piece of paper and places this face down, hands the original to you and returns to their place. Repeat until all children have had a turn. Compare the final number written by each team with the original number.

small pieces of paper, two tables and pencils

Number strings

Ask four or five children to come to the board. Read out a string of four numbers, e.g. 2, 5, 9, 7 and let the children write them. Compare the numbers that the children wrote. Were they the same? Repeat 3 or 4 times. Repeat the activity using strings of 5 numbers then strings of 6 numbers.

Matching numbers

The children can answer the questions on the page. Make up some more questions for the children to answer.

4A, p.1

OBJECTIVE 1.2

Numbers we see

Children discuss with a partner where we might see numbers like these. Some children tell the class about each picture.

4A, p. 2

What numbers mean

new word

Ask the children to imagine that the number shown in each picture on page 2 of the Pupil Book has been **increased** by 1. Discuss what this would mean for each picture, e.g. if the car registration plate was 1 bigger it would belong to a different car, if the number on the shirt was 1 bigger it would mean the shirt would cost more. This way, children will explain what the number in each picture is telling us.

4A, p. 2

Where would you see these numbers?

Write on the board: 10, 500, 67, 18, 32, 006724351, 2 and 974. Children discuss with a partner where each of these numbers might be seen, e.g. a coin, a packet of sugar, a price sign, a calendar, a temperature gauge, an airline ticket, the page number in their Pupil Book or the price of a new television. Choose children to report to the class. Accept any reasonable answers.

Draw about numbers

Children think about where they see numbers and draw a picture about one of these numbers. Display the pictures. Ask the children to bring examples of numbers on food packets or in newspapers etc. from home so that they can add them to the class display.

OBJECTIVE 1.3

Comparing numbers

new word

Write 73 and 68 on the board. Ask which number is **bigger** and how they can tell, e.g. 73 has **more** tens so it is bigger. Choose two children to show the numbers and then compare them. Write more pairs of numbers on the board for children to model and compare.

bundled sticks, base ten blocks, strings of shells or any material that can be used for modelling numbers to 100

Numbers in lines

Children write a number between 100 and 999. Choose two groups of 6 children to come to the front with their numbers and quickly arrange themselves in order, holding their numbers in front of them. Choose two more groups of children to do the same thing. Have the whole class arrange themselves in order in one long line.

a small piece of paper for each child

Less and more game

new word

Have the children hold the number they made in *Numbers in lines* and as you call out instructions, the children have to sit down until there is only one child left. For example 'the numbers **less** than 140 sit down', 'the numbers more than 823 sit down', etc.

Guess the number

One child in a group writes down a 3 digit number. The other children in the group have to guess what the number is. They ask questions such as, 'Is the number more than . . .? or 'Is the number less than. . .?' Discuss good questions to ask. The winner has the next turn.

Ten numbers

4A, p.3

Children discuss with a partner what they see on the page. (Different ways to represent numbers.) They write down the ten numbers and then write them again in order from the smallest to the largest. Then tell the children to answer the question at the bottom of the page.

OBJECTIVE 1.4

Number words concentration

Hold up the number and word cards one by one for the children to read. Place the cards face down. A child selects two, reads them out and shows them to the class. If the cards match they keep them and the next child has their turn. If the cards do not match the child replaces them and the next child has their turn. Continue until all the pairs are found.

10 cards with the numbers 76, 129, 288, 310, 495, 536, 641, 756, 860, 972 and 10 matching word cards

Matching words with numbers

In groups of 5 or 7 children choose pairs of the number and word cards plus one extra, unmatched card. One child deals them to their group. They try to find their partner. When partners are found they show their word card to the rest of the group, to read out. The child with the unmatched card deals the cards next.

the number and word cards from *Number words concentration*

Number housie

Place the number cards along a ledge and tell the children to write 4 of the numbers in words. Remove the cards. Replace one card at a time. When a child sees one of their numbers, they can tick it off. When a child has ticked off all four of their numbers they shout, 'Housie!'.

the number and word cards from *Number words concentration*

Writing number words

4A, p.4

a dice for each group, a counter for each pair

Read through all the words on the page with the children. In groups of four, one pair of children play against another pair. Taking turns to throw the dice, each pair moves their counter along the number path. They write the word from the square they land on. After both pairs have reached the finish they use their words to make the largest number they can.

OBJECTIVE 1.5

Estimating

new word

Spread out 60 shells. Ask children to **estimate** how many shells there are. Ask if there is an easy way to count the shells and listen to their suggestions. If they do not suggest it, show them how to make groups of ten and then count by tens to find the total. Whose **estimate** was the closest? Children can work in pairs to **estimate** and count other collections.

60 shells or similar small objects

Seeds in a jar

new word

Let children **predict** how many seeds are in a jar. In pairs the children discuss ways to estimate and then some tell the class their **strategy**. Try out any **strategy** that does not involve actually counting all the seeds. Work out the answer that each of these methods produces. Count out the seeds by making groups of ten and counting up how many tens there are. Compare this answer with the guesses and the answers reached by the other strategies. Discuss which methods were most accurate, and which methods were quickest and easiest to use.

a jar containing about 200 small objects such as seeds or stones

Outside estimating

Give each group of 4 children a 1 metre string with the ends tied together and have them spread the string out on the ground. They work out how many stones (or blades of grass) are inside the string. Groups should discuss their strategy and try it out. Encourage children to think of quick and easy methods. Compare results and discuss the different strategies used.

a one metre length of string tied together at the ends for each group of 4 children

Estimating groups

Look at the bottletops in Frame one and guess how many there are without counting. Then count the bottletops. Compare the two answers. Discuss how they counted the bottletops: by ones? by fives? by tens? Which method do they think is best? Repeat this procedure for the other frames. Discuss which frame was easiest to estimate and which was easiest to count.

4A, p. 5

UNIT 2

MONEY 1

This topic builds on getting and giving change, making different money amounts, and shopping (see *Teacher's Resource Book* 3, Unit 7).

ENGLISH WORDS/EXPRESSIONS USED

counting on (or up), multiples of ten, total, least, most, round off, nearest, expensive, cheapest

MAIN IDEAS

- **amounts of money can be made in different ways**
- **it is not necessary to give an exact amount of money when paying for goods**
- **more than one item can be bought at a time**
- **there are ways of working out how much more one item costs than another**
- **different items can be bought for the same amount of money**

OBJECTIVES

The children will:

2.1 use coins and notes to make different money amounts
2.2 calculate how much change they will get after buying various items
2.3 use their own methods to calculate total prices of goods
2.4 find the difference between two amounts of money
2.5 work out how to pay for items from a price list

SPECIAL METHODS AND STRATEGIES

- **Counting up (on) to calculate change using the coin denominations, e.g. if an item costs K1.25 and you gave K2, then you can count K1.25, K1.30, K1.50, K2 where you have used 5t, 20t and 50t to reach K2. The children may find it easier to start off by only using 5t and 10t coins, e.g. for the above example count K1.25, K1.30, K1.40, K1.50, K1.60, K1.70, K1.80, K1.90, K2. They should use play money coins when doing this.**
- **Rounding off to the nearest kina. For example, if an item is worth K1. 85 it is nearly K2. This is useful when estimating if you have enough money to buy items.**

OBJECTIVE 2.1

Making K1

Spread out some coins on surfaces. Provide a list of items that costs K1. Allow children to make up K1 using different coins. Each child writes their number sentence on the board. Discuss if there are other ways of making K1.

coins or coin rubbings for 5t, 10t, 20t, and 50t

Making K1 in different ways

Using coin rubbings/play money, children make K1 using two coins, then 3 coins, then 4 coins and then 5 coins. Each time they write a number sentence about the coins they used. Each time compare solutions.

coin rubbings/play money for 5t, 10t, 20t, and 50t

What are notes worth?

Let children look at the notes. They should describe the features of each note. Discuss what each note is worth. Have children use one kina coins to show you the value of each note.

notes for K2, K5, K10 and K20 and K50 if available, one kina coin rubbings

Making amounts of money

Explain to the children that when we write kina and toea, the numbers before the dot are kina and those after the dot are toea. Let children look at the price tags on page 6 of the Pupil Book. Now, read out the instructions on the page. Ask one child to explain the example. Children write at least two different number sentences for each amount.

counters

4A, p. 6

Making K100

Work out how many of each note (K2, K5 and K10) would be needed to make K100.

Calculating change

new word

Set up a 'class shop'with prices less than K1. Revise giving change to one kina. Children choose an item and work out how much change they should receive by **counting on** from the price of the item in tens or **multiples of ten** until they reach K1. For example, if the item is worth 75t and you gave K1 then you can count 75t (give 5t), 80t (give 10t), 90t (give another 10t) to make K1. So 25t is the amount of change to be given.

a collection of items with prices marked under K1, coin rubbings

OBJECTIVE 2.2

Counting money

People who work in banks count money by putting the coins in groups so that each group contains the same coins. Ask children to do this with a handful of coin/rubbings play money, and then count each pile. (Good practice in counting by 1's, 2's, 5's, 10's, 20's and 50's.) They should also count the **total** amount of money they have. They write the amounts in words and figures.

new word

coin rubbings/play money

Class shop

Have a 'shop' set up on a table using items priced more than K1 but less than K10. Work out which item(s) cost the **most** and which item(s) cost the **least**. Ask questions like 'What two items can you buy for less than K10 from this shop?'. Discuss the need to keep records in shops.

new word

shop items priced more than K1 and less than K10

Shopping

Give a child a K2 note and let that child choose an item they want. They give 'a shopkeeper' the K2 who gives them the change. Let the children discuss whether the change is correct and explain how they worked this out. Choose more children to buy an item with K5 and with K10 and let the other children work out how much change should be given. Let children tell how they worked it out.

play money coins and notes to represent K2, K5 and K10

How much change?

Ask children to discuss the pictures in Frame A. Remind the children that the numbers before the dot are kina and those after the dot are toea. (You want to buy items that together cost 5 kina and 50 toea, you give the shopkeeper K10, how much change should you get?) Let children use coin rubbings and 'play notes' to work out how much change they should get and write this amount down. The children do the same for the other items on the page. Check all answers and encourage children to talk about the different ways they used to find an answer.

play money coins and notes to represent K2, K5 and K10

4A, p. 7

OBJECTIVE 2.3

Can I afford more than one?

Choose an item from the class 'shop' between K1 and K2. Tell the children that you have K5 and ask them to estimate if you could buy five of them. Ask children to explain how they worked out their answer. Choose another item from the 'shop' worth just under K2 and tell the children that you have K10. Ask them to estimate if you would be able to buy five. Encourage children to **round off** to the **nearest** kina.

new word

the shop items used for *'Class shop'*

Shopping for more than one

Children work in pairs. They choose an item from the shop and estimate how much two of their items will cost. They then work this out accurately. Allow them to use coin rubbings and play notes if they wish. Let some children explain. Let one pair manage the shop and sell/collect. During the unit let different pairs 'manage'.

the shop items used for *'Class shop'*

How much will it cost?

new word

Look at the prices of the items on the page. Talk about which is the most **expensive** item and which is the least expensive (**cheapest**) item? Read through the questions on the pupil page, and ask the children to write their answers in their books.

4A, p. 8

Will we need more?

Children discuss and make lists of situations where we buy more than one of an item.

Reading labels

Choose items from the class shop. Ask children to find out everything they can from the labels.

OBJECTIVE 2.4

Which is worth more?

Have two children each choose an item from the class shop and say how much it is worth. Ask children which item is worth more. Then work out how much more. Choose two or three children to explain how they worked this out. Repeat.

the shop items used for *'Class shop'*

Comparing prices

Read through the questions with the children. Ask them to write their answers in their books.

4A, p. 9

Investigating prices

Is a two kilogram packet of washing powder twice the price of a one kilogram packet? etc.

How much will we buy?

Think of times when it is better value to buy items in smaller packs. For example, there is no point in buying a large amount of a product if it is not going to keep and you have to throw some of it out.

OBJECTIVE 2.5

What's the total cost?

Ask children to look at page 9. Ask questions about the drawn objects such as, 'How much would I pay for a _______ and a _______?', 'How much would I pay for two ______s, a ______, and a ______?', 'If I spent K20 what could I buy?' etc.

4A, p. 9

What did I buy?

Children in groups should work out the problem, 'I spent exactly K45. What did I buy?' Groups are to work out as many different answers to this as possible. Share the answers. How many ways did groups find?

5 or 6 items costing between K5 and K20 drawn on the board

The cost of goods

Children work out the answers and write them in their books. Make a list on the board of some of the different ways they could have spent the K50. Have children work out what you could buy for K30, K40, or K100.

4A, p. 10

Rounding off prices

Let children use newspaper advertisements to choose a set number of items that they would like to buy. Round off each price to the nearest kina and then work out how much money they would need to buy their items.

toy and electrical goods catalogues, colour specials pages from newspapers

UNIT 3

READING TIME

This topic builds on reading, saying and showing times to the hour and half hour (see *Teacher's Resource Book* 3, Unit 6).

ENGLISH WORDS/EXPRESSIONS USED

time, clock, hour, minute, o'clock, half past, quarter past, quarter to, digital display, seconds

MAIN IDEAS

- **we can show and read the time on digital displays and clocks**
- **each number on an analogue clock represents a time interval of five minutes**
- **we say time differently for digital displays than clocks**
- **time can be shown and read to the nearest minute**
- **there are 60 seconds in one minute and we can use seconds to measure shorter periods of time**

OBJECTIVES

The children will:

3.1 **read, say, and display time to fifteen minutes using clocks and digital displays**
3.2 **read, write, and say time to five minutes**
3.3 **read, write, and say times before the hour**
3.4 **read, write, and say time to one minute**
3.5 **use minutes and seconds to describe and estimate the duration of short events**

SPECIAL METHODS AND STRATEGIES

- **A clock looks like**

- **A digital display looks like**

- **When reading digital time read what the numbers say, e.g. 7:05 is said as 7,0,5; 4:15 is said as 4,15, etc.**

OBJECTIVE 3.1

Saying the time

new word

Use the demonstration clock to revise parts of the **clock**. Carefully demonstrate times including **o'clock**, **half past** and **quarter past** in words, **clock** and **digital** times. Children discuss and show times by moving the clock hands. Emphasise the correct way of saying the **time** and if there is more than one way, read and say the time each way.

a demonstration clock with moveable hands

Reading and writing the time

Show a time on the clock-face (hour, half hour and quarter hour only). Get the children to read the time out loud and write the times shown in their books. Share their answers with the class.
With a partner, one child writes a time, e.g. 2:15 and the other child shows, reads and says the time on the clock-face.

a demonstration clock

Making hands

Get children to make hands for the large clock-face in Frame A. Say a time both clock-face and digital and ask children to show the time on their clock. Check. Repeat.

sticks or card to make hands for the clock

4A, p.11

Clock times

Using the small clocks children write the time shown on each clock-face in Frame B. For each of the digital times shown, children draw a matching clock in Frame C.

4A, p. 11

OBJECTIVE 3.2

To five minutes

new word

Revise that there are 60 **minutes** in 1 **hour**. Using the demonstration clock put the long hand on the 12. Then move the long hand to the 1 and ask the children how many **minutes** the hand has moved. Continue to move the hand around the clock, stopping on each number as the children count by fives until you reach 12 again. Move the hand again and have the children count in fives until you are back at 12. Write digital times of five **minutes** and say the times, e.g. 8:05 (say 8,0,5), 8:10 (say 8,10), 8:15 (say 8,15), etc.

a demonstration clock

Clock-face and digital time to five minutes

Using both clock hands show some times to five minutes on the demonstration clock, e.g. 10 past 7, 25 past 9, quarter past 11. Get the children to discuss and say the times. Ask other children to discuss, write and say the same times as they would see on digital clocks. Groups should report their answers. (Note: Carefully show the hour hand when showing time past the o'clock time.)

a demonstration clock

Showing time to five minutes

Use the long and short hands to show the following times on the big clock, e.g. 5 past 10, 20 past 3, 25 past 1, quarter past 12. Make up a story about each time.

the long and short hands that were made in *Making hands*

4A, p.11

Matching pictures and times

Write the time shown on each clock as both digital and clock-face time. Draw clock-faces to show the three digital times on the page. Children work with a partner to discuss the pictures, say the times and decide which clock shows the time the action in each picture takes place.

4A, p.12

Class time

Draw six clock-faces showing various times to five minutes throughout the school day. Display them in the class room. For each time, draw a picture to show what the class was doing at that time. Match the picture with the time. Do the same activity over a few days. Discuss whether the same things are done each day.

OBJECTIVE 3.3

Times to the hour

new word

Use the clock to show some time to the hour, e.g. 9:50. Ask a child to explain what time this is. Talk about how they know. Show some other times on the clock-face, e.g. 9:45 (**quarter to** 10), 9:40 (20 to 10), 9:35 (25 to 10). Children explain what the time is. Name some activities they do in class at the times shown.

a demonstration clock with moveable hands

To five minutes

Set the demonstration clock at 4 o'clock. Move the minute hand forward in five minute intervals, 5 past 4, 10 past 4, quater past 4, 20 past 4, 25 past 4, . . . , each time asking a child to say what time it is till you reach the half past mark on the clock. Then show the children how the time will now get closer to 5 o'clock than 4 o'clock, so we start saying how long it is 'to'5 o'clock. Continue to move the hand forward 5 minutes at a time, each time asking a child to explain what the time is. When you reach 5 o'clock, continue moving forward by five minute intervals as the children read the time shown until you reach six o'clock, then you stop.

a demonstration clock with moveable hands

Reading digital time

Explain how to read a digital display. When we are reading a digital display we read what the numbers say. Show 6:10 on your digital display and ask a child to read it (6,10). Make the same time on the clock-face and ask a child to read it (10 past 6). Children practise with each other using times *to* and *past* the hour. Use digital display and clock-face times. Check as they work, listening carefully to what they say.

a digital display and a clock-face

Matching digital displays to clock-faces

Children make a list of the ten digital times on paper. They read and say the times. Show each time on a clock-face then write the time beside the digital times, e.g. A) 8:10 = 10 past 8. H) 6:40 = 20 to 7. Check each pair while they are working. Have one child read their answers to the whole class.

the long and short hands that were made in *Making hands*

4A, p. 13

OBJECTIVE 3.4

Sixty minutes in an hour

Ask the children to show minutes and a five minute interval. Children talk about: 'how many equal divisions' (minute marks) between 12 and 1. Count with the children. Talk about what each of the marks represent. Discuss how many minute marks are around the clock-face.

a demonstration clock

Time to one minute

Show some times on the clock, e.g. 3:12, 8 past 5, 6:48, while children say the times shown. In groups, children talk about and show times to one minute while others demonstrate and say times. Check groups. Ask some children to show the times on the demonstration clock.

a demonstration clock

Showing time to one minute

Using the demonstration clock place the short hand on the hour, say 3 o'clock, and use the long hand to show various times, such as 4 past 3, 3:29, etc. Children work in pairs with one using the hands to show a time on the demonstration clock while the other draws a clock-face showing the time. Both children should read and write the time shown. Then change roles. Check while children are working.

a demonstration clock, the long and short hands that were made in *Making hands*

4A, p. 11

Linking clock times with digital times

Children match the clock times with the digital times.

4A, p. 14

Time concentration

Make 12 cards showing clock-face times to the minute. Make a matching set of cards with the corresponding digital time for each clock-face. Use the cards to play *Concentration*. (Place cards face down.) Children turn over two cards at a time. If the cards match, the child keeps them. If they do not match, they are replaced face down where they were before. Play until all the pairs have been found.

OBJECTIVE 3.5

How many seconds in one minute?

new word

Tell children to watch the seconds hand on the clock. Count together as it goes past the **seconds**. Ask how long it takes to go around the clock once. Discuss how many **seconds** are in one minute. Clap hands for one minute as the seconds hand travels around the clock-face. Discuss what happens to the minute hand over one minute and then over five minutes. Talk about what happens to the **seconds** hand in five minutes and the hour hand after one minute.

a large clock with a seconds hand

How long is one minute?

Have all children stand up. From the time the teacher says, 'Now', the children sit down when they think one minute has passed. Time them on your clock. See who is the closest. Discuss how children estimated one minute. Try again using some of their ideas.

Seconds or minutes?

Discuss why we have seconds. Ask children to name some things that take seconds to happen, rather than minutes. Have children predict if standing on one foot would be measured in seconds or minutes. Get them to try this out and check the result. Get them to estimate first. Now check whether the following things would be measured in seconds or minutes: drinking a glass of water, counting to 100, running around the school, etc.

How long?

Children are to read through the list of time words. In pairs discuss how long each activity would take. Children should write down the number for each picture and beside it write the word from the list which best describes how long the activity would take. When they have finished, discuss their answers with the whole class. Children may not agree on some of the answers. It will depend on their culture. Accept reasonable answers and listen to explanations.

4A, p. 15

Time words

Discuss words we use for measuring or counting time, e.g. seconds, minutes, hours, days, weeks, months and years. Children may have heard of decades or centuries. Briefly discuss their meaning. Talk about other words for saying how long something will take e.g. a morning, an afternoon, a season, a moment, blink of an eye, a short time, a long time.

UNIT 4

FRACTIONS 1

This topic builds on recognising and making $\frac{1}{2}$, $\frac{1}{4}$, $\frac{1}{3}$, $\frac{1}{5}$, $\frac{1}{10}$ in symbols and drawings (see *Teacher's Resource Book* 3, Unit 6).

ENGLISH WORDS/EXPRESSIONS USED

model, equivalent, method, mixed number, number sentences, sequence

MAIN IDEAS

- **fractions can be used as labels**
- **fractions can be given in words and symbols**
- **wholes are made up of fractional parts**
- **wholes can be made from different fractions**
- **fractions can be used as counting numbers**

OBJECTIVES

The children will:

4.1 recognise and write common fractions
4.2 make and name common fractions
4.3 explore the relationships between parts and wholes
4.4 explore part-whole relationships with $\frac{1}{2}$, $\frac{1}{4}$, and $\frac{1}{8}$
4.5 count using halves and quarters

SPECIAL METHODS AND STRATEGIES

Children need to say fractions. Make opportunities for oral practice. They also need to write fractions.

OBJECTIVE 4.1

Recognise and write thirds

Choose three children and ask the class several questions about them such as, 'What fraction are girls?', 'What fraction are wearing shorts?'. Ask questions that will produce the answer $\frac{1}{3}$ or $\frac{2}{3}$. Choose children to answer and explain. Write the fraction on the board and explain why the fraction is written this way.

Recognise and write 1/4

Draw and write some different fractions on the board, including $\frac{1}{4}$. Four groups of four children stand up. Ask questions that will produce the answer $\frac{1}{4}$, e. g. 'What fraction of your group is wearing shoes?' and let each group answer for itself. Ask one child to point to $\frac{1}{4}$ on the board. Choose another child to write the fractions on the board as each group answers. Repeat for other questions.

Fraction races

Make two teams. Hold up a fraction card. A child from each team races to be first to write the fraction on the board. The first to do this scores a point. Children explain why the fraction is written this way. Let all children have a turn.

at least 15 cards with diagrams representing a range of unit and other fractions such as $\frac{1}{3}$, $\frac{1}{2}$, $\frac{3}{4}$, $\frac{4}{5}$, etc.

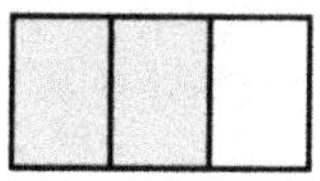

$\frac{2}{3}$

$\frac{3}{4}$

What fraction is this?

In Frame A, the children draw in their books the diagrams which show $\frac{1}{3}$.

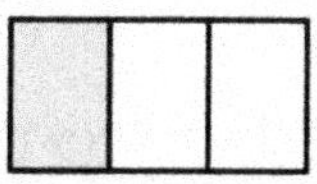

In Frame B, they must look at the diagrams and write the fraction shown in words and symbols.

4A, p. 16

Drawing fractions

Have children draw pictures or diagrams to show a certain fraction, e.g. draw $\frac{3}{5}$ in as many ways as you can.

Fractions in the classroom

Children find examples of fractions in the classroom, e.g. $\frac{3}{4}$ of the windows are open, there is writing on $\frac{2}{3}$ of the board, etc. Write or draw about them.

Fraction 'I Spy'

Start by saying, 'I spy with my little eye $\frac{4}{5}$', and let children look around to find the fraction. The child who guesses correctly has next turn.

Fraction Snap and Fraction Concentration

Give out the fraction cards with matching blank cards. Children write on their blank card the fraction numeral for their fraction diagram. Use these cards to play the games *Concentration* or *Snap*.

the fraction cards used in *Fraction races* and 15 blank cards

OBJECTIVE 4.2

$\frac{1}{4}$ full

Show a bottle with leftover content. Allow children, in pairs, to discuss

1. what fraction of content is left
2. how they could check

new word

Let children explain and demonstrate their **method** for checking. Talk about advantages and disadvantages.

a bottle $\frac{1}{4}$ full of drink, other containers

Fraction tasks

new word

Each task card should ask children to **model** a fraction, e.g. tie a knot $\frac{2}{3}$ of the way along a piece of string, pour $\frac{3}{4}$ of this tub of sand into another container, put a mark at $\frac{3}{5}$ of the height of the doorway, take $\frac{1}{4}$ of the books off our bookshelf. Each group takes a card. Groups use suitable equipment to make their fraction. Let groups show their **model** and explain how they did it. Other children can comment and suggest alternative methods. Display the **models**.

A set of task cards, one for each group. Equipment to go with each task.

Choosing fractions

Children choose the fraction shown for each picture and write it in their books. Discuss the answers. Then they do the question at the bottom of the page.

4A, p. 17

Dice fractions

Let each child throw 2 dice and use the numbers thrown to make a fraction, e.g. 3 and 5 will make $\frac{3}{5}$. Give each child a sheet of paper and tell them to write the fraction, $\frac{3}{5}$, draw a picture about $\frac{3}{5}$ and write a sentence explaining the picture. Check that their fractions are correct. Make finished pictures into a book or a display.

two dice, paper and colouring materials

OBJECTIVE 4.3

Fraction number sentences

new word

Let children watch as you fold a piece of paper and cut it into quarters. Show two of the quarters and ask a child to write how much you show, e.g. $\frac{1}{4} + \frac{1}{4}$. Pick up another quarter and ask a child to continue the number sentence to show what you are holding now, e.g. $\frac{1}{4} + \frac{1}{4} + \frac{1}{4}$. Hold up all four $\frac{1}{4}$'s and ask a child to complete the **number sentence,** e.g. $\frac{1}{4} + \frac{1}{4} + \frac{1}{4} + \frac{1}{4} = 1$. Repeat for other fractions.

pieces of paper

Making fraction strips

new word

Give each child 3 strips. Colour one strip in red and fold it in half. Write $\frac{1}{2}$ on each section, then cut it in half. Fold another strip into quarters and colour the whole strip green. Write $\frac{1}{4}$ on each section and then cut it into quarters. Write 1 on the last strip but do not fold it. Use the cut fraction strips to show **equivalent** fractions, e.g. make 1 using halves and quarters, or how many quarters make a half?

strips of paper about 1 cm wide and 8 cm long, three for each child

Make One Strip

Play in groups of 4. Each child keeps their whole strip and puts their $\frac{1}{2}$'s and $\frac{1}{4}$'s in the middle. They take turns to throw the dice, pick up a strip to match the fraction shown on the dice and place it on their whole strip. Continue like this, adding another fraction each time until they have covered their strip exactly. If the fraction thrown will not fit on their strip they miss that turn. When a child fills their strip they write a number sentence about it, e.g. $\frac{1}{2} + \frac{1}{4} + \frac{1}{4} = 1$. The others check that the sentence is correct. Play the game enough times so that each child writes at least 2 or 3 number sentences. The child with the most correct number sentences at the finish is the winner.

a dice marked $\frac{1}{2}, \frac{1}{2}, \frac{1}{4}, \frac{1}{4}, \frac{1}{4}$ for each group of four, fraction strips

What fraction will finish it?

Children look at the strips being used in a game of *Make One Strip*. Work out what fraction is needed to complete each diagram on page 18 and write it in their books as a number sentence. One example is given.

4A, p. 18

Pattern block fractions

Give each pair of children one of each of the 6 different pattern block pieces. Using the hexagon as one, children should work out what fraction some of the other pieces show. Discuss their answers.

pattern blocks

OBJECTIVE 4.4

Making eighths

Give each child a paper strip and ask them to make $\frac{1}{8}$s. Colour the strip blue, label each $\frac{1}{8}$ and cut them into separate pieces. Put them with the rest of the fraction strips they made and use different combinations of fractions to show 1, e.g. $\frac{1}{8} + \frac{1}{8} + \frac{1}{4} + \frac{1}{2}$.

a paper strip about 1 cm wide and 8 cm long for each child, colouring materials, fraction strips

Play 'Make One Strip' a different way

In groups of four play the game the same way as before, writing down a number sentence for each strip they complete.

a dice marked $\frac{1}{2}$, $\frac{1}{4}$, $\frac{1}{4}$, $\frac{1}{8}$, $\frac{1}{8}$, $\frac{1}{8}$ for each group of four, fraction strips,

Finish the strips

Work out what fraction is needed to complete each strip and then write a number sentence about completed strips. They should use their fraction strips to help them do the other questions.

fraction strips

4A, p. 19

Cover the strips

In groups of four, children put the whole strip in the middle and take turns to place one of their smaller strips along the whole strip until it is covered exactly. Slide these pieces off the strip, leave them in the centre and start making a new strip. If a child only has fractions too big to fit on the strip they miss their turn. The winner of the game is the first child to use all their fraction strips. Once finished, they can write a number sentence to show what they did.

fraction strips for each child and one whole strip for each group

Ordering fractions

Give each child four squares of paper. They are to fold and cut one piece into 2 halves labeling them as halves. They then repeat with a second sheet to make 3 thirds, with the third sheet to make 4 quarters and the last sheet to make 8 eigths, each time being careful to make the parts equal and to label them clearly.

4 paper squares for each child

OBJECTIVE 4.5

Counting by halves

Tell a story about a village where it takes $\frac{1}{2}$ hour to fetch one bucket of water. Write $\frac{1}{2}$ on the board. Ask children how long it would take to make two trips to fetch water. Discuss how they worked this out. Write 1 next to the $\frac{1}{2}$. Continue by asking about 3 trips, 4 trips, 5 trips, etc. each time discussing how children work it out. Keep writing the length of time on the board so you build up the counting **sequence** for $\frac{1}{2}$, e.g. $\frac{1}{2}$, 1, 1$\frac{1}{2}$, 2, 2$\frac{1}{2}$, 3, 3$\frac{1}{2}$, 4 . . . Children should continue the counting **sequence**. Children count out loud, using the written **sequence**.

new word

Counting by quarters

Tell about how it takes $\frac{1}{4}$ of an hour to make a rice holder from vine leaves. Write $\frac{1}{4}$ on the board. Ask children how long it would take to make 2, 3, 4, 5, 6, etc. rice holders each time discussing how they worked it out and building up a counting sequence for quarters, e.g. $\frac{1}{4}$, $\frac{2}{4}$, $\frac{3}{4}$, 1, $1\frac{1}{4}$, $1\frac{2}{4}$, $1\frac{3}{4}$, 2 . . . Children should continue the counting sequence.

Halves on a number line

Draw a large number line on the board and mark in the whole numbers 1, 2, 3, 4, and 5. Ask several children to show $\frac{1}{2}$ on this number line. Discuss their answers. Ask children to mark other halves, e.g. $2\frac{1}{2}$, $4\frac{1}{2}$, etc. Children can draw a number line in their books and continue it up to 8 marking in the halves. Ask the children to read out what they write, in English.

Quarters on a number line

Have children draw a number line from 0 to 5 in their books. Ask them to show $\frac{1}{4}$ and give reasons for their choice. Mark in the other quarters on the number line. Make sure the marks are evenly spaced.

What fraction is it?

new word

Children work out the fraction or **mixed number** marked on each number line and write the answer in their books. Then they should draw a number line for each of the fractions or mixed numbers listed and mark each fraction or mixed number on its number line. In groups, they can read the fractions or mixed numbers aloud.

4A, p. 20

Fraction count

Put the card for $\frac{1}{3}$ on a ledge and give the other cards to 9 children. Ask the children to come out and place their card on the ledge if they think it is the next number in the sequence. Continue until all cards have been placed. Read through the numbers. Repeat, giving the cards to different children. Repeat, but remove 3 of the cards before giving them out. Can the children identify the missing numbers?

10 large fraction cards in a sequence, e.g. $\frac{1}{3}$, $\frac{2}{3}$, 1, 1 and $\frac{1}{3}$, . . .

Fractions in the world

Children can work with a partner to make a list of all the places where fractions are used, e.g. $\frac{1}{2}$ past for clock time, in recipes to give quantities, etc.

UNIT 5

ADDITION

This topic builds on single digit additions, solving story problems, and mental strategies such as adding 10 (see *Teacher's Resource Book* 3 Units 4 and 9).

ENGLISH WORDS/EXPRESSIONS USED

odd, even, grid, add, horizontally, vertically, row, column, diagonal, exchange, units, tens, addition, straight

MAIN IDEAS

- **when adding single digits together we use various strategies to make the task easier**
- **place value is important when adding numbers together**
- **different pairs or groups of numbers can have the same total**
- **a written algorithm is used when adding larger numbers**
- **additions can be written as vertical algorithms or in story format**

OBJECTIVES

The children will:

5.1 add two or more single digits and discuss strategies for adding single digits
5.2 practise adding two or more single digits
5.3 add 1- and 2-digit numbers using materials and symbols
5.4 add 2- and 3-digit numbers using materials and symbols
5.5 solve story problems using addition

SPECIAL METHODS AND STRATEGIES

- **Many activities in this unit are sequential. Choose from the early activities before attempting later ones.**
- **Strategies for adding mentally: when adding numbers in your head look for pairs of numbers that add up to 10. For example, to do 3 + 26 + 7 you would add 3 and 7 first and then add the 10 to 26.**
- **To add 2-digit numbers or larger numbers together add the units first, then the tens, then the hundreds as follows:**

$$\begin{array}{r} {}^{1}4\,5 \\ +2\,8 \\ \hline 7\,3 \\ \hline \end{array}$$

- **Encourage children having difficulty to use concrete materials**

SUGGESTED ACTIVITIES

OBJECTIVE 5.1

Adding odd and even

new word

Investigate what happens if you **add** an **even** single digit to an even single digit; an **odd** single digit to an **odd** single digit; and an **odd** single digit to an **even** single digit or vice versa.

Addition grid

+	5	2	9	6
7				
4				
1				
8				

new word

The children copy a **grid** like this into their books and fill in the spaces by adding each of the numbers across the top with each of the numbers down the side. Time how long it takes children to do this. Ask children to discuss then report on which numbers they found the easiest to add and why, e.g. it is easy to add 1 because you just have to count on 1, or 5 + 7 is easy because 5 + 5 = 10 and then you add 2 more to make 12.

Making 10

Write an example on the board such as 3 + 7 + 6. Ask groups of children to work out the answer and talk about how they found it. Have each group report the strategy they used to find the answer. Write on the board 3 + 26 + 7 and ask the children to work out the answer in their heads. When they have done so they discuss with their group how they found their answer and decide whose strategy is easiest. Let each group report this strategy. Repeat for 6 + 8 + 4 and 6 + 28 + 4.

Adding single digits

Children are to write answers to Frames A and B in their books. Discuss how they did each part.

4A, p. 21

Magic grids

Ask children to copy the grid in Frame C into their books and add the numbers **horizontally** and **vertically** and write the totals for each line. Ask them to add the four **horizontal** answers together and the four **vertical** answers together and to put the total for each in the circle at the bottom. Ask them to explain what they have found (they should both add to 80).

4A, p. 21

Make your own grid

Children can make up their own magic grids for sixteen numbers and add them both horizontally and vertically and check to see if they add to the same number.

OBJECTIVE 5.2

Card circle totals

Jumble each set of card circles in separate piles face down. Two children to compete. They take the top card from each pile and hold them up. The first child to call out the total of the two cards is the winner and sits down. Choose a new child to replace the winner and continue. Continue until every child has had a turn. (If a child 'loses' three times in a row, tell both children to sit down and choose two new competitors.)

two sets of cardboard circles with the numbers 0 to 10

2-digit number sentences

new word

Place one set of the card circles in the centre of the children so all can see the numbers. Write on the board, _ + _ = _. Ask a child to choose three circles to make the **number sentence** true, e.g. 4 + 5 = 9, 1 + 3 = 4, etc. Each circle can only be used once in each sentence. As a true **number sentence** is made write it on the board and put the circles back in the centre so they can be used again for another **number sentence**. Continue until children cannot find any new **number sentences**. Do this once as a class then the children can do it in groups.

one set of cardboard circles with the numbers 0 to 10

Number sentences for 12

Write on the board _ + _ + _ = 12. Ask a child to choose three of the circles to make this sentence true, e.g. 5 + 3 + 4, 7 + 4 + 1, 8 + 4 + 0 etc. Write the sentence on the board. Choose another child to find a different true number sentence for 12 and write it on the board. Children can only use a number once in each sentence. Continue until children cannot find any new number sentences.

one set of cardboard circles with the numbers 0 to 10

Make 14

Give each pair of children eleven small pieces of card. They write the numbers 0 to 10 on these and then use them to find number sentences for 14 using three different digits. Write each one down. Choose some pairs to tell you their answers and write them on the board so that you have a list of number sentences each of which totals 14.

11 small pieces of card for each pair of children

Magic squares

new word

Children should copy the magic squares into their books and complete each one so that each **row**, **column** and **diagonal** has the same total as shown in the circle above the squares. Copy the next two number squares into their books and work out the missing numbers by using the numbers around the outside to help them.

4A, p. 22

Who has the higher total?

Pairs of children shuffle the cards and place them face down. In turn pick up three cards each and add the numbers together. The child with the highest total scores one point. Continue playing until one child scores ten points.

small number cards from *Make 14*

OBJECTIVE 5.3

First to 100

new word →

Each group takes some single sticks and bundled sticks and a dice. One child rolls the dice and takes that number of single sticks. In turn other children do the same. When a child has ten or more single sticks they **exchange** them for one bundle. Continue playing until one child has ten bundles (one hundred). The first child to get one hundred is the winner.

single sticks and bundled sticks, a dice for each group

How many children?

new word →

Tell a story about 32 children in a class who were visited by 9 children from another class. Have groups of children use single sticks and bundled sticks to show 32 and 9. Write 32 on the board and 9 underneath it. Groups use sticks to work out how many children were in the room altogether. Choose a group that added the **units** first, and then added the **tens** to explain and show how they did it. Accept and praise any methods that work. Show the children how this looks on the board, e.g.

$$\begin{array}{r} {}^{1}32 \\ \underline{+9} \\ \underline{41} \end{array}$$

Add the numbers vertically relating it to what the children did with the sticks.

single sticks and bundled sticks

How many children now?

Tell another story about 27 children in a class who were visited by 15 children from another school. In groups, using sticks show 27, then 15. Write 27 on the board and 15 underneath it. Groups put the sticks together to work out how many children were in the room altogether. Choose one group to show how they did it. Do this with the numbers on the board first showing how when the units are added we have one ten and two units just as we had one bundle of sticks and two single sticks. Show how we add the tens together to make four tens. Finally read the answer (42).

single sticks and bundled sticks

Find the answers

Talk about how the example at the top of the page has been done. Tell children that this is the usual way of setting out the sums. Children copy the sums from Frame A into their books and then use materials to work out the answers. If any children want to do this without materials let them do so. Work through the first few and choose different children to show how they worked them out.

sticks, place value squares

4A, p. 23

Find the missing numbers

Ask children to work out what the missing numbers in Frame B might be. They should use materials to help them if they want to. Choose children to tell their answers.

sticks, place value squares

4A, p. 23

Make your own sentence

Have children make up their own missing number sentences for others to work out.

OBJECTIVE 5.4

First to 250

Play the game *First to* 250. Give each group two sets of cards numbered 0 to 10. Shuffle the cards and place them face down in a pile. Each child writes the number 150 in their book. Take turns to pick up a card from the top of the pile. Add the number picked up to the 150 and write the new number in their book. Continue in this way adding each of the numbers they pick up to the last number written in their book. When the pile of cards is finished shuffle them and keep playing until one player reaches 250. This player is the winner.

2 sets of cards numbered 0 to 10 for each group of 4

How many coconuts?

Tell a story about collecting 235 coconuts one day and 145 coconuts the next day. Ask groups to use place value squares or sticks to represent the numbers. Write the numbers on the board. Groups use the material to find how many coconuts were collected altogether. Choose a group who used place value squares to show how they worked out the answer (380). Choose a group who used sticks to show how they did it. On the board show how it is written making sure to relate each step to the materials.

sticks, place value squares

How much money?

Give each group some coin rubbings and have them show 235 and 145. Discuss how they do this. Groups work out the answer and write it as kina and toea (K3.80). Talk about how it is different from writing it as 380.

coin rubbings, e.g. kina, 10 toea and 1 toea coins

Adding 2 and 3 digit numbers

Tell a story about there being 135 children at one school, 67 at another school, and 96 children at another school. Groups represent the numbers and work out the total number of children at schools (298). Do the sum on the board relating each step to the materials.

sticks, place value squares

More addition

Children copy the sums into their books and work out the answers. Use materials to help them if they want to. Work through the sums by choosing a different child each time to tell and explain their answer.

sticks, place value squares

4A, p. 24

Make the number

Each child writes a number of their choice between 10 and 50 on a piece of paper. Call out a number, e.g. 75. Children have to find and join with another child or group of children so that their numbers when added together make 75 or a number close to 75, e.g. one child may have 25, another may have 30, and another may have 20 so that together they total 75. Children sit with their group when they have made their total. When a few groups are sitting down stop the game and ask these groups to show their numbers so everyone can check them. Keep playing by calling out different numbers.

small pieces of paper

OBJECTIVE 5.5

Make up a story

Write on the board, 68 + 17 + 146. In pairs children make up a story about the sum. Let some children tell their story. Children write the sum vertically in their books and work out the answer using materials if they want. Ask them to say the answer to the question as it relates to their story. Repeat with two more sets of numbers.

How many students?

Write on the board, There are 187 students at one school and 56 students at another school. How many students is this altogether? In pairs, children discuss how to work out the answer. They copy the sum into their books and use materials, if they want, to find the answer. Check that children understand what to do. Choose pairs to read out their answer as it relates to the question, e.g. There are 243 students altogether.

Story problems

In pairs children read and discuss the first question. They write the sum in their books and work out the answer. Do each question in the same way. Choose a different pair each time to tell and explain their answer.

4A, p. 25

An addition story

new word

Give each child a page with an **addition** sum at the top, e.g. 76 + 89 = . . . 245 + 57 = . . . etc. Children write a story underneath their sum, work out the answer and write it as part of the story. Make a class **addition** book from the completed pages.

sheets of paper

Adding three numbers

Children copy this diagram from the board into their books.

5	6	4	8	2	9
1	5	3	7	0	6
8	4	4	9	2	3
3	9	0	5	7	8

new word

They find 'strings' of three numbers in a **straight** row either horizontally, vertically or diagonally that add up to 15. Put a line through them. Find as many 'strings' as you can.

How many did you eat?

Tell a story about two brothers who ate 112 peanuts between them. The older brother ate more than the younger brother. Children in pairs discuss how to work this out. Choose some pairs to tell you how they might do it. Accept any suitable methods. Let pairs do it using materials if needed. When children have an answer they write it as an addition sum and check that it does add up to 112. Write some of the sums on the board.

Travelling by bus

Discuss what the chart in Frame A tells us. Ask questions about the chart. In pairs children read the questions underneath the chart and talk about how they will do them. Check that children understand what to do. Let children use materials if they want. Check as they work.

sticks, place value squares

4A, p. 26

Work them out

Children read the stories in Frame B and talk about how to find the answers. They do them and write their answers as addition sums. Check.

4A, p. 26

Class stories

Write some addition sums about your class, e.g. It takes 9 steps to reach the door from my seat and another 15 steps to reach the fence, so altogether it takes 24 steps to reach the fence (9 + 15 = 24).

UNIT 6

CHANCE

ENGLISH WORDS/EXPRESSIONS USED

will, won't and might and a range of other words that indicate chance such as: maybe, possible, probably, perhaps, always, sure, never, outcome, certain, even chance, impossible, likely, fair

MAIN IDEAS

- chance can be thought of as ranging from impossible through increasing degrees of probability to certain
- we have many words to describe the chance of an event occurring
- for any event, there may be a number of possible outcomes
- the chance of a particular outcome occurring depends on how many other outcomes are also possible
- recording the outcomes of a repeated event in a table can help us to compare how often each outcome occurs

OBJECTIVES

The children will:

6.1 use the English words 'will', 'won't' and 'might' in the context of everyday stories

6.2 use a range of English words which describe the chance of something happening

6.3 examine some situations to list all of the possible outcomes

6.4 perform one action a number of times and record the outcomes

6.5 play a dice game and do an activity to explain the outcome

SPECIAL METHODS AND STRATEGIES

This may be the first time the children have met these words in English. Let them practise these words in English stories and sentences. They should listen to, say, read and write these words in complete and meaningful English sentences.

OBJECTIVE 6.1

Chance sentences

new word

Point to each word card in turn. Say the word, and have the children say the word. Make up a sample sentence using each of the words. For example, 'The sun **will** rise tomorrow', 'I **won't** grow to be 3 metres tall', and 'It **might** rain'. Ask children to make up sentences. Check that they use the words correctly.

three cards: one each for the words **'will'**, **'won't'** and **'might'**

Chance/lists

In groups, children talk about and then list five things that *will happen*, five things that *won't happen* and five things that *might happen*. (Children should soon realise that something that will happen for them may/will/won't happen for others and that they need to find things that are true for all members of their group.) Ask each group to read their 'will happen' list. Let other children agree or disagree and say why. Repeat for the other lists.

Chance stories

Tell the children some stories, e.g. 'I will have my fiftieth birthday this year', 'My uncle is travelling to Australia in the next month', 'I am going straight home after school', etc. Ask the children to write one of the three words (*will*, *won't*, or *might*) in their books which best describes what possibility this is for them. Discuss the reasons why not everyone has the same word written for every story.

Will, Won't or Might

Allow children to talk about which of these things will/won't/might happen tomorrow. Discuss their answers. You could also ask children to draw a picture about an event that will/won't/might happen.

4A, p. 27

Predicting the weather

Have a chart displayed with the headings, 'Will rain', 'Won't rain', and 'Might rain'. Let children predict the likelihood of rain for that day by writing their name in a column. At the end of the day see who predicted correctly. Do this each day for one week.

OBJECTIVE 6.2

Chance words

Using the word on each card, children read the word then try to explain what it means. They then make up sentences for each word. Discuss whether the word has been used properly.

cards with words written on them such as: might, maybe, possible, probably, perhaps, certain, will, always, sure, won't, never, no hope, impossible.

Sorting chance words

new word

Give each group of 4 enough small cards. Tell them to copy the chance words onto them and sort their words into the following categories; **certain**, **even chance**, or **impossible**. Let each group show what they have done. Do any of the groups disagree? Allow them to explain/discuss.

a supply of small word cards

Using chance word in context

Ask children to think about situations they have been in where words such as 'maybe' or 'perhaps' have meant 'certain', and other situations where they have meant 'less than an even chance'. Let some children tell about these situations.

Which is more likely?

Ask children to discuss with a partner each of the activities or events shown and describe how likely it is for each to happen. Compare each pair of activities to say which is more likely. (For some children neither of the activities or events will be likely, e.g. having a bigger brother or sister. Discuss why neither is likely for these children.)

4A p. 28

Making predictions based on chance

Make three class charts of things you think will happen in the next week / next month / before the end of the year. As each thing happens place a cross next to it. At the end of the week, month, and year look to see if your predictions were correct.

Sorting chance statements

In turn children take a card, read it out loud, place it in one of the boxes and say why they chose that box.

cards with statements written on them such as, 'I like eating peanuts', 'I can run as fast as a car', three boxes labelled 'certain', 'uncertain', and 'impossible'

OBJECTIVE 6.3

Listing and comparing outcomes

new word

Check that there are the same number of each colour counter. Mix them up and ask the children to close their eyes and pick up two counters, putting one counter in each hand. Ask them to look in their hands to see what colour counters they picked up and put themselves in groups with other children who have exactly the same counters in their hands. Record all the possible **outcomes** on the blackboard, e.g. if the colours were red and green the possibilities would be: 2 red, 2 green; 1 red and 1 green; 1 green and 1 red. Count how many children had each of these possibilities and record this. Discuss.

2 equal sets of different coloured counters, enough for at least two per child

More listing and comparing

Check that there are the same number of each colour counter. Mix them up. Children close their eyes and pick up any two counters, putting one in each hand. Open their eyes, check the colours then group themselves with other children who have exactly the same colours. Record all possibilities. If the colours were red, green and blue the possibilities are: 2 red; 2 green; 2 blue; 1 red and 1 green; 1 green and 1 red; 1 green and 1 blue; 1 blue and 1 green; 1 blue and 1 red; 1 red and 1 blue. Count how many children had each of these possibilities and write it on the board. Discuss.

3 equal sets of different coloured counters, enough for at least two per child

Listing possible outcomes

Choose different children to tell what they think might happen next, to the children in the first box. List all the possibilities on the blackboard. Discuss if some of the possibilities are more likely to happen than others. Repeat for the other two drawings. Look at the first picture in Frame B. Let children work in pairs to make a list of the possibilities for the order of age of families with two children (2 boys, 2 girls, 1 boy and 1 girl, 1 girl and 1 boy). Some children may find it easier to use two different colour counters to represent boys and girls. Now look at the second picture of a family with three children. With a partner list all the possibilities of this family. Again children may find it easier to use counters and to then list them in columns, e.g.

boy, boy, boy (red, red, red)	boy, girl, boy (red, green, red)
girl, girl, girl (green, green, green)	girl, girl, boy (green, green, red)
boy, boy, girl (red, red, green)	girl, boy, boy (green, red, red)
boy, girl, girl (red, green, green)	girl, boy, girl (green, red, green)

Ask some pairs to read out their list and record them on the board.

4A, p. 29

OBJECTIVE 6.4

Heads and Tails game

Use the coins shown in Frame A to make sure that the children know which side of each coin is heads and which is tails. (The 'head' is the side with the PNG crest and the 'tail' is the side with the animal.) Use any of the words that you use locally, e.g. 'kumul' and 'muruk' or 'kumul' and 'kapul'. Before you toss the two coins, each child must predict what the result will be by either putting both hands on their head (two heads), or both hands on their tail (two tails), or one hand on their head and one hand on their tail (one head and one tail). After you have tossed the coin, the children whose prediction is not correct sit down. Continue playing until one person, the winner, is left standing.

two coins

4A, p. 30

Heads or tails

Discuss the table drawn in Frame B. Ask, 'How many times was the coin tossed?', 'How many times did it land on heads?' and 'How many times did it land on tails?' Now tell them to draw a similar table then toss their coin twenty times and to record what happens. Draw a similar table on the blackboard and record the totals for each pair of children. Work out how many tosses there were altogether and how many of these were heads and how many were tails.

a coin for each pair of children

4A, p. 30

Two coin toss

Tell one child from each pair to toss the two coins together twenty times and the other child to record what happens on a table like the one shown in Frame C. Draw a similar table on the board and when pairs have finished record all the outcomes on the class table. Discuss the results, particularly the fact that the number of 'head/tail' combinations is approximately equal to twice the number of either two heads or two tails.

two coins for each pair of children

4A, p. 30

OBJECTIVE 6.5

Teacher against the class

Explain that two dice will be rolled and the two numbers uppermost on the dice added together. If the total is 2, 3, 4, 10, 11, or 12 the class scores one point, and if the total is 5, 6, 7, 8, or 9 the teacher scores one point. Tell the children that they have six numbers while you only have five. Keep a tally on the board. Play the game by letting children take turns to roll the dice and calculate the sum. Continue until either the teacher or the class score 20. Discuss the results. Is the game **fair**? Give reasons.

new word

two dice

Recording outcomes

Ask children to make a copy of the chart shown in Frame A. Each pair rolls their two dice 100 times, adds the two numbers thrown, records each throw on their chart, then finds the totals. Collect the results and display on a class chart. Discuss the results. Ask why 7 occurred much more frequently than 2 or 12, why 1 did not occur at all etc. Then ask children what five numbers they would choose if they were going to play the game again.

two dice for each pair of children

4A, p. 31

Is it fair?

Look at Frame B. The chart shows the results of a game of *Teacher Against the Class* where the teacher's numbers were 6, 7, 8, and 9, and the class's numbers were 2, 3, 4, 5, 10, 11, and 12. Ask, 'Did this class have a better chance than the first game we played?', 'Is this a fairer way to play the game?', etc. If there is time play the game using these numbers and compare the results with the results shown on this page.

4A, p. 31

UNIT 7

MENTAL CALCULATIONS

This topic builds on all the whole number work the children have done so far, with special emphasis on using strategies for mental calculations (*see Teacher's Resource Book* 3, Unit 9).

ENGLISH WORDS/EXPRESSIONS USED

double, halve, subtract, difference between, multiply, divide, method

MAIN IDEAS

- **there are a number of 'short-cuts' that we can use to calculate mentally**
- **these 'short-cuts' can often be used to help us deal with larger numbers**
- **there is usually more than one way to solve a number problem**
- **it is useful for children to hear what strategy others used to solve a problem**
- **children need to be able to use number bonds flexibly in a variety of situations**

OBJECTIVES

The children will:

7.1 **add and subtract 9 and 99, practise doubling and halving and multiplying by 10**

7.2 **use mental strategies to solve number sentences**

7.3 **use their knowledge of number bonds to 20 to create magic squares**

7.4 **see that various mental strategies can be used to help them solve problems with larger numbers**

SPECIAL METHODS AND STRATEGIES

A key to this unit is to let the children work out strategies for themselves and to discuss their strategies with each other. Accept all strategies which work. Expect some children to have difficulty in explaining their strategies. However, they should be encouraged to talk about their strategies in whatever language they are most comfortable with, as often as possible.

OBJECTIVE 7.1

Adding nine

Place the cards face down in the centre. A child picks up a card and adds 9 to the number on the card. If they are correct, they keep the card, if they are not correct, they have one more try. Keep playing until all the cards are gone. During the game, ask some children to talk about how they add 9. An easy way of doing this is to add 10 and **subtract** 1. However, accept other efficient methods.

new word

a set of twenty cards, each with a number between 400 and 600

Subtracting nine

Play the *Adding Nine* game but this time ask the children to subtract 9 from the number on the card. During the game, ask some children to talk about how they subtract 9. An easy way of doing this is to subtract 10 and add 1. However, accept other efficient methods.

a set of twenty cards, each with a number between 400 and 600

Adding ninety-nine

Choose a child to throw the three dice and say what number is represented, e.g

= 451.

Then they add 99 to that number. During the game, ask some children to talk about how they add 99. An easy way of doing this is to add 100 and then subtract 1. However, accept other efficient methods.

3 dice—one is the hundreds dice, one is the tens dice and one is a units dice

Subtracting ninety-nine

Play the *Adding ninety-nine* game but this time ask the children to subtract 99 from the number on the dice. During the game, ask some children to talk about how they subtract 99. An easy way of doing this is to subtract 100 and then add 1. However, accept other efficient methods.

3 dice—one is the hundreds dice, one is the tens dice and one is a units dice

Double the dice

In turn, each child throws the dice, then **doubles** the number thrown and writes it down. Continue doing this, adding each new score to the previous score until you reach one hundred. First to reach one hundred is the winner. After the game talk about how children **double**. Some will **multiply** by two and others will add the number to itself. (Both ways are suitable.)

new word

a dice for each group of four children

Halving numbers

Ask pairs of children to each write down six numbers which come between 200 and 900. Tell pairs to discuss how they could find half of each of their numbers. Talk about the ways the children suggest and try them out to see which are successful. Some children will **divide** by two and others will subtract the number from itself. How can you check if your way works? (Double or multiply by 2 to see if you reach your starting number.) You could also talk about what happens when you **halve** an odd number.

new word

Multiplying by 10

a strip of paper for each child

(You might like to practise the 10 times table before you start this activity.) Children take a strip of paper and write on it ten different numbers less than 100, one under the other, e.g. 67 48 91 85 13. . . Then they swap strips of paper with someone else. Holding the strip of paper down the side of a page of their workbook they write in their book, next to each number on the list, the number that is made by multiplying the number on the strip by 10. (The strips can be swapped around and used several times.)

Number noughts and crosses

Instead of using noughts and crosses for this game you use the numbers from 1 to 9. Each number may be used only once in a game. Play with a partner and take turns writing a number in a space on the grid. Draw some grids in your workbook. The idea is to be the one to get the numbers in any row, column or diagonal to add up to 15, e.g.

2 5 9
4 3 7
8 6 1

After children have had two or three games, discuss strategies they can use to help them win the game. You may need to introduce a rule that stops children from putting 5 in the middle.

Making into tens

Write the following addition on the board: 8 + 2 + 7 + 3 + 5 + 5 + 6 + 4 + 9 = . . . Ask the children to add these up in their heads. When a few children have worked out the shortcut of grouping numbers into tens let them explain their method to the rest of the class, e.g.

$$\underbrace{8 + 2}_{10} + \underbrace{7 + 3}_{10} + \underbrace{5 + 5}_{10} + \underbrace{6 + 4}_{10} + 9 = \mathbf{49}$$

Write several more examples making the combinations a little less obvious each time, e.g.

$$7 + \overbrace{5 + 3 + 8 + 5}^{10} + 9 + 2 + \overbrace{4 + 6}^{10} + 1 + 8 = \mathbf{58}$$

(below: 7 + 5 + 3 = 10; 8 + 5 + 9 + 2 = 10; 9 + 2 + 4 + 6 + 1 = 10)

Making up to 100

Tell the children that the trade store owner has to make up packets of 100 biscuits. Each packet already has some biscuits in it. The first packet has 57 biscuits, how many more are needed to make 100? Invite the children to give an answer and to explain how they worked it out. If no child uses the counting on method explain how to add biscuits to make them up to the next 10, e.g. 57 and 3 makes 60, then 60 and another 40 makes 100. So 43 biscuits are needed altogether. Repeat this for 48, 75, 62 and 39.

OBJECTIVE 7.2

Stand Until You Don't Know

Slowly call out a list of numbers less than 10, e.g. 5, 3, 2, 6, 4 . . . You will have to write the list on a piece of paper so that you can check the answer. The children must add the numbers in their heads until they lose track of the total. When this happens they sit down. The last child standing, who knows the correct total, is the winner. Play two or three times.

Mental Quiz

Tell the children to work out the answers to the questions in their heads. They are to write down just the answers. Read out the first set of ten questions. When you have finished, correct the children's answers. Repeat with two more sets of ten questions.

3 sets of 10 questions that the children can calculate in their heads, e.g. 35 + 7, 5 x 3, 40 - 3, 16/4, 2 + 4 + 6, half of 20. . . .

Calculating in your head

Tell the children that they are going to work out the answers to all these sums in their heads. Ask them to say the answer to the first sum in Frame A. Then ask the children to say how they worked it out. (Don't accept answers like 'I just knew'. Make them explain that they doubled the number or they added 20 plus 20 then 5 plus 5 etc.) Now ask them to work out the answer to the next question (51) and explain how they worked it out. Make sure they relate it to the first question, e.g. '26 is 1 more than 25 so I just added 1 more to 50.' Repeat for the other questions, making sure that each time they relate the answer to the previous questions. Repeat this procedure for the other Frames on the page.

4A, p. 32

OBJECTIVE 7.3

Identifying Magic Squares

6	11	4
5	7	9
10	3	8

4	9	2
3	5	7
8	1	6

Tell the children to look closely at the numbers in the first magic square to try to work out why it is called 'magic'. When they have worked out that all the columns and rows and the diagonals add up to 21, let them work out the sum of rows, columns, diagonals on the other magic square. Discuss whether the same thing happens as in the first.

two magic squares drawn on the board

Creating Magic Squares

new word

Children to work with a partner to draw a 3 x 3 grid in their books and try to create another square that is 'magic'. If no-one has succeeded in doing this after 5 minutes provide the children with a clue. Say, 'Look at the two magic squares on the board. What is the **difference between** the numbers in the upper left hand corner of each square?'(2) Now ask, 'What is the **difference between** the numbers in the centre of each square?'(2) Let the children continue to investigate until they find that the **difference between** numbers in corresponding positions is always 2. Let them use this fact to create a new version of the magic square. Tell the children to copy their magic square onto the card but to leave out 4 of the numbers. Let the children swap cards with someone nearby then copy the square into their books and 'solve' it by writing in the missing numbers. They should let their partner check that their solution is correct.

small pieces of card, about 15 cm x 15 cm, enough for one per child

Magic Squares

Tell the children to copy the smaller squares into their books and work out the missing numbers. When they have done this they should investigate the large square at the bottom to see if it is a magic square. When most children have finished, choose some to read out their solutions to the magic squares and also to explain whether the large square at the bottom of the page is 'magic' or not.

4A, p. 33

OBJECTIVE 7.4

Discussing mental strategies

Write 36 + 25 = . . . on the board. Tell the children you are interested in the different ways you can work out the answer in your head. Ask the children for the answer (61) and write it on the board. Now ask for volunteers to explain the methods they used to work out this answer. For example, 30 and 20 are 50 and 6 and 5 are 11, 50 and 11 are 61; double 25 is 50 and 36 has an extra 11 so the answer is 61. As the children explain their methods be careful not to rate them by responding with terms like 'That's fantastic!' or 'Yes, it's OK to do it that way.' If the method works and the child is happy with it, then it is all right. Repeat the procedure with:

7 × 16 =
86 − 47 =
K3.45 + K2.67 =
58 + 38 =

Using mental strategies

4A, p. 34

Children should work out the answer for each number sentence in their heads without writing anything down. If, after 5 minutes, there are still children who have not worked out the first answer, stop the class and invite some children to give their answer and to explain the method they used to work it out. (For example, 'We added them together so they made groups of 10 and then we added on the last number.') When most children have finished, choose some to read their answer for each question and explain the method they used.

UNIT 8

SUBTRACTION

This topic builds on methods of subtraction including subtracting with zeros, finding the difference between numbers, and solving story problems (see *Teacher's Resource Book* 3, Unit 5).

ENGLISH WORDS/EXPRESSIONS USED

take away, trade, pattern, half way between

MAIN IDEAS

- **people use different strategies for subtracting mentally**
- **we can record subtractions vertically to show the trading processes involved**
- **finding the difference between two numbers involves using subtraction**
- **finding half way between two numbers involves using subtraction**
- **subtraction can help us solve many everyday problems**

OBJECTIVES

The children will:

8.1 revise taking away using mental strategies
8.2 taking away and finding the difference between two numbers
8.3 explore finding numbers half way between two other numbers
8.4 perform subtraction tasks that involve zeros
8.5 use subtraction to solve story problems

SPECIAL METHODS AND STRATEGIES

The usual way to set out subtractions is like this:

Step 1

$$\begin{array}{rrr} 5 & {}^{6}\not{7} & {}^{1}2 \\ -\,1 & 8 & 5 \\ \hline & & 7 \end{array}$$

Step 2

$$\begin{array}{rrr} {}^{4}\not{5} & {}^{16}\not{7} & {}^{1}2 \\ -\,1 & 8 & 5 \\ \hline & 8 & 7 \end{array}$$

Step 3

$$\begin{array}{rrr} {}^{4}\not{5} & {}^{16}\not{7} & {}^{1}2 \\ -\,1 & 8 & 5 \\ \hline 3 & 8 & 7 \end{array}$$

OBJECTIVE 8.1

Take Away 100

new word

Children play the game in pairs. They each make 100 with their material. Then, taking turns, they say aloud what number they have in front of them, roll a dice and **take away** the number shown. If there are not enough ones to take away the required amount, they **trade** a ten for ten ones and then take away the number showing. The game continues until one child, the winner, reaches zero.

material for modelling numbers: base ten longs and cubes, bundles of sticks and loose shells, one dice for each pair of children

Sharing strategies for taking away

Write 63 – 7 on the blackboard. Ask the children to work out the answer in their heads and then to write down their answer. Choose several children to tell the strategy they used to work out their answer. (Most will say they took 3 from 63 to make it 60, that left 4 out of the 7, so they took away the 4 which left 56. Some may say they took away 10 which left 53, but this was 3 more than the 7, so they added on another 3 to make 56.) Repeat with three more subtractions: 83 - 8, 34 – 9 and 62 – 5.

Taking away in your head

Tell children to work out the answers without doing any working out on paper, just in their heads, and write the answers in their books. When enough children have finished, choose a different child to give their answer for each of the questions and explain how they worked it out. Let other children who used different methods tell the class how they got their answer.

4A, p. 35

Last one standing

Ask the children to stand. Tell them to start at the number 50 and take away each number that is called out. Anyone who cannot work out the answer in their head must sit down. Call out numbers such as 5, 2, 3, 6 . . . until there is only one child left or you reach zero. Do this several times.

OBJECTIVE 8.2

Modelling and recording subtraction

Tell the children a story: a plantation has 615 timber trees. The villagers worked very hard for three days and cut down 270 of the trees. Now they want to work out how many trees are left standing. Write 615 – 270 (vertically) on the blackboard. Ask one child to show 615 trees using the material. Now ask how we can take away the 'trees' that have been cut down so that we can see how many

base ten blocks, bundles of sticks, strings of shells or other material for modelling numbers are left. Make sure the child giving the explanation starts with the units, then explains how one of the hundreds must be traded for 10 tens so the 7 tens can be subtracted, and then subtracts the hundreds. After each step show, on the blackboard, how the process is recorded with numbers.

Doing subtraction

Tell children that in the box at the top is the subtraction recorded when two boys wanted to work out how many coconuts one of them had collected. They knew they had collected 362 coconuts altogether and one of the boys had counted his 175 coconuts as he collected them. So they knew if they took away 175 from their total they would find out how many the other boy collected. Tell the children to make 362 with their materials and let them do the taking away with their materials step by step, as you discuss how the process is recorded with numbers. Then ask the children to copy out the rest of the subtractions on the page and to use their materials, as well as numbers, to work out the answers. Choose different children to give their answers and explain how they worked them out.

base ten blocks, bundles of sticks, strings of shells or other material for modelling numbers

4A, p. 36

Subtraction patterns

new word

Write this subtraction on the blackboard: 120 – __ = __. Ask the children to work in pairs to find two numbers that will make this number sentence correct. When they have done this, ask them if there might be other pairs of numbers that would also be correct. Give them time to work out what some of these would be. When they have had enough time, start writing some of the answers on the blackboard. Can the children see any **pattern** in the answers? Can they work out how many different answers are possible? (120, e.g. 120– 0 = 120, 120– 1 = 119, 120 – 2 = 118, 120 – 3 = 117, 120 – 4 = 116 . . .

Finding the difference between

Label one of the mountains Mt Dremsel, Manus Province, 725 metres, and the other Mt Eunainaun, New Ireland, 657 metres. Ask the children how they could work out the difference in the heights of the two mountains. Remind the children that finding the difference between 2 numbers means we take away the smaller number from the larger number to see how much is left over. That amount is the difference between the two. Choose one child to show the height of Mt Dremsel using the base ten blocks or other material. Choose another child to show how to take away the height of Mt.Eunainaun Choose a third child to show how to write this process in numbers as a subtraction.

a picture of 2 mountains on the blackboard, base ten blocks, bundles of sticks, strings of shells or other material for modelling numbers

How much would you save?

Tell the children that Joe is having a sale in his furniture store. Ask questions to make sure they understand the difference between the usual price of goods and their sale price, e.g. 'How much does a television usually cost?', 'What is the sale price of the refrigerator?'. Work out the difference between the usual price and the sale price on the radio, showing children how to write this down as a subtraction. Tell them to write each of the other questions as a subtraction and then work out the answers, using their material to help them if they need to. Choose children to give their answers to the questions and explain how they worked them out. Take time to discuss the different methods of working out the answers to the other questions on the page.

materials for modelling numbers

4A, p. 37

Electrical Goods Sale

Make up some prices about a sale of TVs and stereos marked down from K863 to K699, from K945 to K799, from K776 to K599, etc. Have the children work these out in their heads and discuss what methods they used.

OBJECTIVE 8.3

Strategies for finding half way between

new word

Ask the children how to find the number exactly **half way between** 10 and 70. Let them work in groups of four then ask each group to explain their strategy very carefully. Praise those groups who use a picture or diagram. A number line is a very good way of demonstrating '**half way between**'. Repeat, this time finding the number **half way between** 25 and 55. Now pose this problem to the children, 'When I was four years old I weighed 15 kilograms and when I was eight years old I weighed 35 kilograms. How heavy do you think I was when I was six years old?'Ask them to discuss in their groups ways of working out the answer. Let each group report back to the class.

Finding half way between

In Frame A children must find the number exactly halfway between the two numbers shown on the number line. In Frame B they must work out what price comes exactly half way between the prices shown for the different items. In Frame C they must find the weight that is half way between each pair of items shown. In Frame D they must find out what length pencil is half way between the two pencils shown. When the children have finished their work, choose a different child each time to explain their answer to the class.

4A, p. 38

OBJECTIVE 8.4

Modelling subtractions involving zero

Ask the children to model 107 with their materials. Ask what is unusual about the model of this number. (It has no tens.) Choose a child to explain how the number 107 tells us that it has no tens. Now tell the children that you want them to take ten away from 107. Write on the blackboard:

$$\begin{array}{r} 1\,0\,7 \\ -\quad 1\,0 \end{array}$$

Ask how this can be done? Ask them to discuss it in their groups, and then try it out using their materials. When they have finished choose several children from different groups to explain what they did. (Changed their hundred into 10 tens and then took one of the tens away.) Now ask a child to come to the blackboard and show how this can be recorded in numbers, e.g.

$$\begin{array}{r} \not{1}\,{}^{1}0\,7 \\ -\quad 1\,0 \\ \hline 9\,7 \end{array}$$

Repeat, this time using 150 − 4 = Choose a child to come to the blackboard and show how this can be recorded in numbers, e.g.

$$\begin{array}{r} 1\,{}^{4}\not{5}\,{}^{1}0 \\ -\qquad 4 \\ \hline 1\,4\,6 \end{array}$$

base ten blocks or other materials for modelling numbers

Doing subtractions involving zero

Work out the two examples in the box at the top of the page together. Tell the children to make 360 with their materials then go through the process of subtraction one step at a time, starting with the units. After each step show the children how the subtraction is recorded in numbers. Repeat this for the second subtraction in the box at the top of the page. Tell the children to copy down the other examples and use their materials to help them work out the answers. Encourage the children to explain the process they use to work out the answers each time. Each time they should record the subtraction in numbers, not just write in the answer. Choose different children to come to the blackboard to show how they worked out their answers using the numbers.

4A, p. 39

Examples involving hundreds

Work through this special example with the children in the same way as in the previous activity:

$$\begin{array}{r} {}^{2}3\ {}^{9}0\ {}^{1}0 \\ -\quad 4\ \ 3 \\ \hline 2\ \ 5\ \ 7 \end{array}$$

Give them two or three money problems to work out on their own such as 'How much change would I get from K5 if I spent 75t?'

OBJECTIVE 8.5

How many passengers?

Tell the children a story about an aeroplane with 52 passengers flying from Goroka to Port Moresby. On the way it landed at Lae and 47 passengers got off. How could the hostess find out how many passengers were still on board without counting each one? Write 52 - 47 (vertically) on the blackboard and tell the children to solve the problem with their materials and then to write it down and work it out using numbers. Ask one child to explain the method they used and show how to record it using numbers. Emphasise the links between the materials and the numbers. Make up a similar subtraction story and repeat the above steps making sure that all children are confident with the process.

materials for modelling numbers

Solving story problems

Children read each problem and then find the answer by modelling it with their materials. Then they should write the problem down in numbers and solve it in writing. Choose different children to explain how they got their answers and to show their working out on the blackboard.

4A, p. 40

Take away book

Give each pair of children one sheet of paper. Show them how to fold it into thirds. In the top third they are to write 2 numbers between 100 and 900. Then all the children swap sheets so each pair has a sheet from another pair of children. In the middle third of the paper they write a take away story using the 2 numbers written in the top third of the page. When they have finished they all swap papers with a different child, and then they record the take away story as a subtraction sum in the bottom third of the page and work out the answer. Collect all the sheets and make them into a Take Away Book.

UNIT 9

ANGLES AND DIRECTION

This topic builds on understandings of turns and part turns, compass directions, and giving and following directions (see *Teacher's Resource Book* 3, Unit 2).

ENGLISH WORDS/EXPRESSIONS USED

turn, quarter turn, half turn, left, right, amount of turn, angle, angle tester, right angle, direction, tangram, compass, compass points

MAIN IDEAS

- **angle is the amount of turning around a point**
- **a right angle is equivalent to quarter of a full turn**
- **angles can be compared and ordered by placing one over the other or by using an angle tester**
- **direction can be described in terms of points on a compass**
- **we can give directions or follow them using the compass points**

OBJECTIVES

The children will:

9.1 take part in activities which show an angle as an amount of turn

9.2 be introduced to the word 'angle' and use an angle tester to find 'right' or 'quarter turn' angles

9.3 estimate whether angles are bigger or smaller than a 'right' or 'quarter turn' angle

9.4 compare the size of angles

9.5 use compass points as a way of describing and following directions

9.6 play a game which revises compass points

9.7 be introduced to the compass points NE, SE, SW, and NW to describe and follow directions

SPECIAL METHODS AND STRATEGIES

This unit is more about thinking, talking and listening than writing in English. Give the children plenty of opportunities to talk in English. The children should already know these concepts from earlier grades. This unit reinforces these concepts using the English language.

OBJECTIVE 9.1

Left and right turns

new word

All the children should face in the same direction. Give instructions like: make a **quarter turn** to the **right**, make a **half turn** to the **left**. Check that all children are able to follow the directions in English.

Simon says

Give directions for children to follow. If you do not say, 'Simon says', the children must not do it or they are 'out'. Some possible directions are: 'Turn one quarter to the right', 'Take three steps backwards', 'Do a whole turn', 'Make a half turn to the left'. Keep the games short.

Action patterns

Put children in groups of four to make a repetitive 'dance' or 'jump' pattern using quarter turns, half turns and whole turns. Let them practise before performing them for the other children. They can make a clapping pattern to go with their movement. When the children are confident, make one child the 'caller' of the movements and the whole grade could participate.

Which way to turn?

Ask questions about the picture such as, 'If the person standing makes a quarter of a turn right, then another quarter of a turn right what will they be facing?', 'If the person makes a quarter of a turn right, and then half a turn left, what will they be facing?' Let children work in pairs asking each other similar questions in English.

4A, p. 41

OBJECTIVE 9.2

Turns on a clock-face

new word

Have both hands on the clock pointing to 12. Move the minute hand from 12 to 3. Discuss how far it has turned (a quarter turn). Move another quarter of a turn. Talk about the **amount** of turn (a half turn). Continue discussing the amount of turn until the hand has turned a full circle.

a demonstration clock

Introducing the term angle

new word

Using the clock-face, show children how the amount of turn becomes larger as the minute hand moves a quarter of a turn and how it becomes larger still when it moves half a turn, then three quarters of a turn, and lastly a full turn. Tell the children that this 'amount of turn' is called an **angle**. Clearly demonstrate which **angle** you are talking about. Write the word **angle** on the blackboard.

Making an angle tester

Have each child make an angle tester (see page 000) by joining their strips together at one end with the fastener. (If these are not available you could use a drawing pin or a nail and put foam or cork on the sharp point once it has been pushed through the cardboard.)

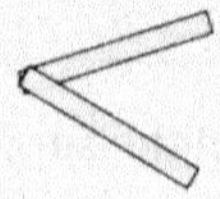

two thin strips of cardboard and a paper fastener for each child

Right angles in the classroom

new word

Demonstrate how to make an angle with their angle tester. Have them make an angle to show a quarter turn, e.g. (A/W 9.2ii) and tell them that this 'quarter turn' angle is called a '**right angle**'. Let children 'fix' their angle testers in this position and use them to find some more 'right' or 'quarter turn' angles in their room, e.g. corners of books, corners of tables, the angle where the wall meets the floor, etc.

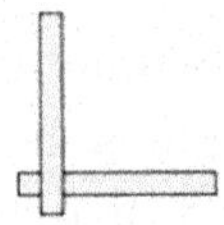

Making a right angle

Hold up an angle tester which has the arms together and tell children that you need to move the arms of the tester so that they show a 'right' or 'quarter turn' angle. Do this and then ask the children to suggest ways they can check if they have done it accurately, e.g. hold it against something you know is a right angle such as a book corner.

Finding right angles

Tell the children to use their angle tester to find the 'right' or 'quarter turn' angles. They can also find those angles bigger than a right angle and those angles smaller than a right angle.

4A, p. 42

Right angles on the clock-face

Find some clock times where the position of the two hands makes a 'right' or 'quarter turn' angle. Accept any reasonable answers, e.g. 5 past 4.

OBJECTIVE 9.3

Drawing and testing angles

Have children draw five angles they can see which they think are bigger than a 'right' or 'quarter turn' angle. Draw five angles they can see which they think are smaller than a 'right' or 'quarter turn' angle. When they have finished remind children how they can check that their angle tester is exactly a right angle and then use their angle tester to check if the angles they chose were correct.

angle testers

Looking at angles

Discuss the shapes drawn on the page by giving instructions and asking questions such as, 'Point to the triangle', 'Which shape has no angles?', Have children follow the instructions on the page.

4A, p. 43

More or less than a right angle?

In pairs, children talk and write about the angles in the shapes on the page which they estimate to be bigger than a right angle and less than a right angle.

4A, p. 43

about 10 counters for each pair of children

Angles in the classroom

Find other angles inside and outside their classroom which are equal to a right angle, bigger than a right angle or smaller than a right angle. Make a class list of these.

angle testers

OBJECTIVE 9.4

Angles in a tangram

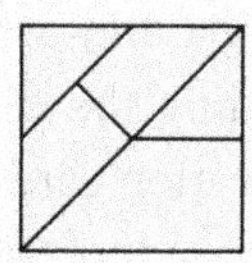

new word

Tell the children their task is to work out which of their five **tangram** pieces has the biggest angle. Do not tell them how to do it. Let each group tell their answer and say how they worked it out. Ask each group to hold up a shape with a right angle. Are there other shapes with right angles? (Yes) How many right angles are there altogether? Repeat by asking them to find out how many angles there are that are smaller than a right angle, then how many angles there are larger than a right angle.

a five piece tangram for each group of three children

Angles in the classroom

Ask the groups to find the biggest and smallest angles in the classroom, using their angle testers to help them compare angles.

an angle tester for each group of children

Ordering angles

Tell the children to do things such as put a counter on all the angles that look like right angles, or put a counter on all the angles that are bigger than a quarter turn. Then tell children to put the angles in order, recording the letter sequence in their books. They can use their angle testers to help them. Let children share their answers.

counters, angle testers

4A, p. 44

Angles on a clock-face

Find some times when the arms of a clock make less than a quarter turn.

Materials: a clock-face

OBJECTIVE 9.5

Direction game

Display the small direction name cards on the appropriate walls of the classroom. Show the children the large direction card.

North South East West

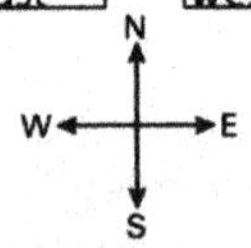

Call out a direction for the children to face. Do this several times. When they can do this confidently, remove the cards and ask questions such as, 'What object can you see that is South?', 'Name two things you can see that are West', etc.

Materials: a large direction card, small direction name card

Directions in the playground

Ask children to copy the 4 directions from the large card on to their small card. Take them outside and ask them to identify east. One way of doing this is by finding the direction where the sun rises (this is east). Place their direction card so that the word 'East' faces the east and then ask, 'What direction is opposite East?'(West). Now look where North and South are and identify some trees or buildings that are in each of these directions.

Materials: large direction card, a small cardboard square for each pair of children to make their own direction card

Hidden Treasure

Write directions such as, 'From the classroom door, take 20 steps, then turn right one quarter turn.'Take the children to the starting point and put the directions where all children can see them. In pairs children are to follow the directions to find the treasure. The first pair to find it gets to keep it.

Materials: a suitable 'treasure' hidden in the school ground, e.g. a 'medal', a 'ribbon', etc., directions for finding 'treasure' written on a large piece of paper

Island journey

Ask questions about the map such as, 'What direction would you be walking in if you were travelling from the hut to the cave?', 'What direction would you be walking in if you travelled from the ship to the captain's house?', etc. Also ask questions like, 'I am going East, where am I walking to?'. Accept all correct answers.

4A, p. 45

Treasure Hunts

Children can make their own treasure hunts (like in Hidden Treasure) for others to follow.

OBJECTIVE 9.6

Directions

With children standing and facing one direction (example, South), give directions to them such as, 'Turn to the east', 'Turn to the west', 'How far did you turn when you turned from east to west?' (half turn), 'Make a quarter turn to the right. What direction are you facing?' (north), etc.

Grids

new word

Make sure that children know how to use a grid by asking them to put their finger on C3. Check. Repeat for two or three more grid references. Choose some children to give grid references for others to find too. Ask children to put their finger on C4. Now ask them to move one space north. They can refer to the **compass points** at the top of the page. Continue giving directions until you have checked that children are using the **compass points** correctly.

4A, p. 46

Pirate game

Explain the rules of the game: each child has their own grid in front of them. One child from each pair puts a counter in a space on their grid but does not show the other child. The other child tries to guess where the hidden treasure (the counter) is. If they do not guess the first time they are then given another clue by the first child. For example, the first child puts the counter on A2 and the second child guesses D3. The first child says, 'No' and gives another clue such as, 'Go east and south'. The first child then has another guess. If they are still wrong they are given another clue. This continues until they guess where the 'hidden treasure' is. Swap roles and repeat the activity.

a counter for each pair of children

4A, p. 46

OBJECTIVE 9.7

Introducing NE, NW, SE and SW

new word

Draw a large **compass** like this **(A/W 9.7i)** on the ground and ask children to help you label the 4 points. Then add two more lines to the **compass** like this **(A/W9.7ii)**

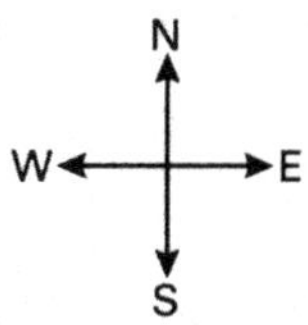

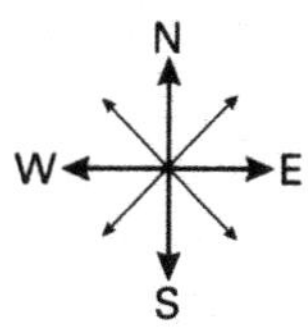

Point to NE and tell children that this direction is called North East. Ask them if they can explain why this might be. Then ask them to predict what the names of the other directions might be. Label the **compass** as you go.

What's in that direction?

Choose children to stand in the centre of the compass and tell you what they can see, for example in the North West. Repeat for other directions.

Simon Says

Play Simon Says using instructions based on the eight compass directions.

Directions on a map

Discuss what the map shows and what the compass means. Use the map to ask questions such as how do I get to the play area if I am in the small garden?

4A, p. 47

Playground Trails

Let children use the eight compass directions to plan a trail around the school. Let other children try to follow it.

Making a compass

Tell children to make a cross on their card and to mark in the four main compass points. Now tell them to add two more lines over these and mark in NE, NW, SE and SW.

a small piece of card for each child

Finding directions

Choose a child to stand in the centre of the compass and tell what direction a particular tree or building is in. Ask the child to align their own compass with the large one and to see if the building or tree is in the same direction. Repeat with different objects for some other children. Now draw a new compass in another part of the playground and label the points. Ask children what direction the tree/building is in now. Discuss why the various features of the playground are now in a different direction from where you are standing.

the large compass from *Introducing NE, NW, SE and SW*, small compasses for each child

Journey to Rock Island

Discuss with the children the directions the boat will sail to get through the islands. Children write in their books the set of directions the boat will take to get to Rock Island. Choose a child to read out their directions while the other children check.

4A, p. 48

Place Names

Find and list places in PNG that have compass directions in their names, e.g. West New Britain Province, Eastern Highlands.

a map or atlas

UNIT 10

MULTIPLICATION 1

This topic builds on multiplication as groups of, and the formal process of multiplying (see Teacher's Resource Book 3, Unit 8.

ENGLISH WORDS/EXPRESSIONS USED

number line, multiples, product

MAIN IDEAS

- **six times table—children should have many ways of remembering new tables facts, including building on number facts they already know**
- **tables facts—are essential for dealing with larger multiplication tasks**
- **setting out multiplication vertically—helps keep track of all the steps involved**
- **multiplication by 10—is necessary before children can multiply by 2 digit numbers in the next multiplication sub-topic**

OBJECTIVES

The children will:

10.1 hear and say the facts of the six times table

10.2 play games which will revise and use the 2, 3, 4, 5, 6, and 10 times tables

10.3 solve some multiplication problems

10.4 use formal strategies for multiplying numbers

10.5 see multiplication by 10 as an extension of place value

SPECIAL METHODS AND STRATEGIES

Short multiplication

$$\begin{array}{r} {}^{1}3\ 4 \\ \times\ 3 \\ \hline 1\ 0\ 2 \end{array}$$

OBJECTIVE 10.1

A line of sixes

at least 60 counters, number cards for the multiples of 6, e.g. 6, 12, . . . 60

Children make out a line of six counters and then find the correct number card for this number of objects (6). They place it at the end of the line. They add another six counters and add the card which shows the new total of objects now in the line (12). They continue until the line has 60 counters altogether and all the correct number cards, like this: iiiiii [6] iiiiii [12] iiiiii [18] iiiiii [24] iiiiii [30] iiiiii [36] iiiiii [42] iiiiii [48] iiiiii [54] iiiiii [60].
As a class, count by sixes up to 60 several times.

Writing the 6 times table

Using this line of sixes as a model, or using any other materials the children can construct the 6 times table. On the blackboard write 1 × 6 = . . . and then ask a child to point to the part of the number line that tells us the answer. Write in this answer. Now write 2 × 6 = . . . etc. underneath the 1 × 6 = . . . until you have written all the number facts for the six times table up to 10 × 6 = 60.

Patterns in the 6 times table

Ask the children if they see any patterns. (The numbers 1 . . . 10 down the left hand side, if you look at the final digit in each answer you find the first five digits are repeated, e.g. 6, 2, 8, 4, 0, all the answers are even numbers etc.) Together, read all of the number facts for the six times table. Allow children to discuss any patterns.

Learning the 6 times table

Now rub out the answers to the multiplication facts, leaving just the questions, e.g. 1 × 6 = ... , 2 × 6 = ... , 3 × 6 = . . . etc. Tell the children to copy these into their books and then work out the answers and write them in. They may look at the number line on the floor if they want to. In groups, they go through the table facts one by one, discussing how they worked out each answer. (1 × 6 is obvious; 2 x 6 is 6 + 6 or double 6; 3 × 6, 4 × 6, 5 × 6 and 10 × 6 should be remembered from the 3, 4, 5 and 10 times tables that they learnt last year. This means that only four of the number facts 6 × 6, 7 × 6, 8 × 6 and 9 × 6 should be new facts to remember and these can be derived from the 3 times table, e.g. 7 × 3 = 21 so 7 × 6 will be double this which is 42)

Which table fact?

In pairs, children discuss the pictures and write the table fact shown in each picture. When they have finished, choose a child to read out their answers.

4A, p. 49

Practising the 6 times table

Practise the 6 times table whenever you can.

Hold up the cards for the multiples of six one at a time and tell the children to write down the tables fact for that number, e.g. you hold up 24 and the children write 6 × 4.

In groups, the children take turns to throw a dice, multiply the number thrown by 6 and add the answer to their tally. The first to reach a score of 200 wins.

Remove one of the cards for the multiples of six from the pack and lay the rest face up in the centre of a circle. See who can be first to say which card is missing. (The children must count by sixes to work this out.)

small number cards for the multiples of 6, e.g. 6, 12, . . . 60

OBJECTIVE 10.2

500

new word

Children throw the dice, multiply the two numbers together and write the answer (called the product) in their book. On each turn they add the new **product** to the previous one. The aim is to be the first to reach 500.

two dice for each group of four children

Grid games

Children draw a grid like the one in Frame A, and choose eight of the numbers from Frame A to write in the squares of their grid. Write a multiplication number fact on the blackboard, e.g. 5 × 6, making sure the answer is in Frame A. The children cover the answer if it is in their grid. Repeat with different multiplication facts. The first child to completely cover their grid is the winner.

eight small clean items, e.g. seeds, for each child

4A, p. 50

Three in a row

Each player has to choose one number from each of the clouds in Frame B, say them out loud, write them down and multiply them. When players have an answer they must find it on the grid and put a counter on it. The first player with three counters in a row horizontally, vertically or diagonally is the winner.

eight counters for each child, a different colour than their partner's

4A, p. 50

OBJECTIVE 10.3

Tables facts for 36

Say, 'Last night I multiplied two numbers together and got the answer 36. However I have forgotten what the two numbers were. Can you help me work out what they might have been?' Let children work with a partner and write down as many pairs of two numbers as possible (2 × 18, 3 × 12, 4 × 9, 6 × 6). Repeat for the number 48 (2 × 24, 3 × 16, 4 × 12, 6 × 8).

Fact Frame

Write - - × - = on the board. Tell the children to fill in the blanks with the numbers 1, 2 and 3 to make as many different number sentences as possible. Then, work out the answers. Some possible answers include 12 × 3 = 36; 32 × 1 = 32; 21 × 3 = 63

Make this true

Write the following on the board.

3–
×2
– –

Children work out the missing numbers to make this true. How many combinations can they find? They can discuss their answers.

Four in a row

Children play the game, following the rules supplied. One dice and some counters are needed for each group. Four cards of each of the following are needed for each group: ×2 ×3 ×4 ×5 ×6 ×10

4A, p. 51

OBJECTIVE 10.4

Recording multiplication

Tell the children the story about the company that had 3 buses that could each seat 32 people. Ask them to help you work out how many people the buses could carry altogether. First of all choose a child to write a number sentence about the buses on the blackboard both vertically and horizontally. Now ask another child to use materials to show the three lots of 32 people (3 lots of 3 tens and 2 units). Check. Tell the children that when we do multiplication like this we work out the units first, then the tens. Count up the 3 lots of 2 units and show the children where to record the 6 in the number sentences on the blackboard. Say, 'Now let's multiply the tens. 3 lots of 3 tens equals 9 tens altogether. Then show how to record this on the blackboard. Make sure that the children understand how the numbers relate to the materials. Relate the answer back to the problem by saying, 'There are 96 people in the 3 buses.' Repeat with more non-trading problems.

materials for modelling numbers, 2 or 3 kinds if you have them

Multiplication with trading

Say, 'The trade store owner had three sacks of rice each containing 25 kilograms of rice. How many kilograms of rice was this altogether?' When they multiply the units to get 15, show them how 10 of the units are traded for a ten and the remaining units are recorded as part of the answer. This ten should then be added on to the other six tens to make seven tens altogether. Make sure that the children understand how the numbers recorded on the blackboard relate to the materials. Relate the answer back to the problem by saying, 'There are 75 kilograms of rice in the 3 bags.' Repeat with more trading problems.

materials for modelling numbers, 2 or 3 kinds if you have them

More trading

Tell a story like, 'There are 4 lines of people waiting at the airport . Each line has 24 people in it. How many people are there altogether?' Have pairs of children show the 4 lines of 24 people with their materials, and then write the number sentence vertically. Check. Write it on the blackboard. Ask each pair to put the 3 lots of 4 together and to trade the ten units for a ten. Now ask them to put all the tens together. Say, '4 lots of 2 tens plus 1 more ten equals 9 tens.'Give pairs time to work out the answer (96). Show the children how this relates to the number sentence by having one group demonstrate the equipment while you write the numbers. Relate it back to the problem by saying, 'There are 96 people waiting at the airport.'

materials for modelling numbers, 2 or 3 kinds if you have them

Solving multiplication problems

In Frame A, children discuss how each of the multiplications were worked out. As the children give their explanations, they can draw the objects in their books to demonstrate what they are saying. The children then use the materials to model the vertical number sentences in Frame B and write the answer to each one. Check.

materials for modelling numbers, 2 or 3 kinds if you have them

4A, p. 52

Multiplication past 100

Ask the children to model 32 x 4 and work out the answer. Choose one child to record what happens on the blackboard. Discuss what happens when you multiply the 3 tens by 4. How do you record this? Try some more multiplications where the answer is greater than 100.

materials for modelling numbers, 2 or 3 kinds if you have them

OBJECTIVE 10.5

Modelling multiplication

Write 10 × 3 on the blackboard and ask the children to model this. Check and write in the answer. Now write 3 × 10 on the blackboard and ask the children to model this. When they have done it, write in the answer. Discuss what ways these number sentences are similar and in what ways they are different.

material for modelling numbers: base ten blocks, 1t and 10t coins and place value squares are all suitable

Looking for patterns

Tell the children to show 5 with their material, then ask them to show you ten times this many. Watch how they work this out. Choose several children to explain what they did and what their answer is. Ask what is the easiest way to make 10 times a given number? (Swap each of the units for a ten.) Write 5 x 10 = 50 on the board underneath the two number sentences already there. Can the children see a pattern? Repeat, this time starting with 7. By now, most of the children should see that when we multiply a number by 10, we swap each of the units for a ten and so end up with an answer that has no units and the number we started with has moved over into the tens place. Some children may see this as simply 'adding zero' to the number they started with.

material for modelling numbers

Multiplying by 10

Children may use their materials to help them work out the answers for Frame A and Frame B but most of the class will find this unnecessary. When they have finished, correct the answers together. How many different answers did they find for Frame B? Discuss how many different answers are possible.

4A, p. 53

Team Races

Hold up one of the cards, e.g. 47. The first member of each team must run to the board and write the result of multiplying the number by 10, e.g. 470. The one who has written a correct answer and is first to return to their place scores a point for their team.

a set of cards of numbers between 11 and 50

UNIT 11

FRACTIONS 2

This topic builds on making and naming fractions, part/whole relationships, and counting in halves and quarters from Unit 4.

ENGLISH WORDS/EXPRESSIONS USED

No new English words. Reinforce the words learnt in Unit 4 Fractions 1.

MAIN IDEAS

- **mixed numbers—fractions combined with whole numbers, e.g. $3\frac{2}{5}$**
- **comparing fractions—is more complex than comparing whole numbers because the size of the fraction decreases as the denominator increases**
- **equivalence—some fractions, with unlike denominators, are equivalent to each other, e.g. $\frac{2}{3} = \frac{4}{6}$**
- **modelling fractions—they can be represented as part of lengths, shapes or groups**
- **fractions are part of our everyday surroundings**

OBJECTIVES

The children will:

11.1 recognise and write mixed numbers
11.2 use fraction walls to compare fractions
11.3 use fraction walls to find equivalent fractions
11.4 will identify fractions as part of groups

SPECIAL METHODS AND STRATEGIES

It is important that children can see and use a model of fractions as a first step. The materials suggested in this unit are important.

OBJECTIVE 11.1

Wholes and quarters

Cut one of the oranges into quarters. Remove one quarter. Discuss what is left. First, write how much of the cut orange is left on the board ($\frac{3}{4}$). Now ask how we can write how many oranges there are altogether. Establish that we write this number of oranges as $3\frac{3}{4}$. Explain that $3\frac{3}{4}$ is a mixed number because it is made up of a whole number and a fraction. Continue cutting up and removing quarters, each time choosing a child to write the mixed number for the oranges remaining.

4 oranges a knife, and a large serving plate

Wholes and fifths

Repeat the previous activity using the bananas, this time cutting them into fifths.

5 bananas, a knife, and a large serving plate

Wholes and eighths

Repeat the previous activity using the mangoes, this time cutting them into eighths.

4 mangoes, a chopping board, a

Recognising mixed numbers

The children look at the picture to find examples of mixed numbers. They write each example they find in their books, e.g. There are $3\frac{1}{2}$ jars of lollies.

4A, p. 54

OBJECTIVE 11.2

Making a fraction wall

Show the children one of the strips and pin this first strip to the wall, then take another strip and ask a child to fold it to show halves. Ask the child to explain what they have done. Pin this strip to the wall directly underneath the first strip and label the two $\frac{1}{2}$ s. Repeat this procedure, using the other strips, for $\frac{1}{3}$s, $\frac{1}{4}$s, $\frac{1}{6}$ths and $\frac{1}{8}$ths. Explain that you have already done a strip to show $\frac{1}{5}$ths and $\frac{1}{7}$ths because it is hard to get the folds exactly right for these. (Look at the fraction wall on page 55 of the Pupil Book, if you are not sure.) Pin all the strips into position to form a complete fraction wall. Leave this fraction wall on display for the children to refer to.

8 large strips of paper, at least 50 cm x 8 cm and a large sheet of paper or a wall to pin them on. Have one of the strips folded into 5ths and another folded into 7ths.

Patterns in the fraction wall

In groups the children can discuss any patterns they notice in the wall. (The pieces get smaller on each line, the denominators get bigger each time, the vertical line down the centre.)

Comparing unit fractions

Ask the children to find $\frac{1}{3}$ and $\frac{1}{5}$ on the fraction wall and decide which is bigger. ($\frac{1}{3}$) Ask someone to explain why this is. (The $\frac{1}{3}$ strip has been folded into only 3 equal pieces, the $\frac{1}{5}$ strip is folded in 5 equal pieces so the pieces are smaller.) Choose several more pairs of fractions to compare in the same way.

Which fraction is bigger?

Tell the children to write their answers to the questions in their own books, using the fraction wall to help them work out the answers.

4A, p. 55

Card games

Use these cards for the children to play comparing and ordering games such as the following:

In turn four children take a card from the top of the pile and place it face up in front of them. The child with the highest/lowest fraction scores a point. First to have 5 points wins.

Each child is dealt 5 cards face down. On 'Go' the children turn over their cards and put them in order from largest to smallest. First to finish wins.

a set of 20 unit fraction cards for each group consisting of: 4 each of $\frac{1}{2}, \frac{1}{4}, \frac{1}{3}, \frac{1}{5}$ and $\frac{1}{8}$

OBJECTIVE 11.3

Equivalent fractions on the fraction wall

Give a child one of the fraction pieces labelled $\frac{1}{3}$. Ask them to say what the fraction is and find a piece the same size on the fraction wall. Now ask the child if there are other fractions on the wall the same size as $\frac{1}{3}$. If the child points to the other two $\frac{1}{3}$ sections, praise this, but tell them to keep looking, until they come to $\frac{1}{3}$. Ask if anyone can write a number sentence about this on the board. ($\frac{1}{3}=\frac{2}{6}$) Repeat this by giving another child $\frac{1}{4}$ and asking them to find another fraction of the wall the same as $\frac{1}{4}$ ($\frac{2}{8}$). Write the number sentence $\frac{1}{4} = \frac{2}{8}$ on the board. Then repeat again.

the fraction wall used in *Making a fraction wall* and 8 large strips of paper, cut into fractions to match the fraction wall, and labelled

Finding equivalents for 1/2

Hold up $\frac{1}{2}$ and ask the children to work out with a partner any other fractions that are the same size as this and write them in their books. When they have written their answers, invite children to give an answer and then use the fraction pieces and the fraction wall to demonstrate it. Build up a number sentence $\frac{1}{2} = \frac{2}{4} = \frac{3}{6} = \frac{4}{8}$.

the fraction wall used in *Making a fraction wall* and 8 large strips of paper, cut into fractions to match the fraction wall, and labelled

Finding equivalent fractions

Hold up two of the $\frac{1}{8}$ths side by side. Ask the children to find a fraction the same as this on the wall. When they have found $\frac{1}{4}$ ask one of the children to record $\frac{2}{8} = \frac{1}{4}$. Repeat with two $\frac{1}{6}$ths, four $\frac{1}{8}$ths, three $\frac{1}{4}$s and two $\frac{1}{3}$s.

the fraction wall used in *Making a fraction wall* and 8 large strips of paper, cut into fractions to match the fraction wall, and labelled

Comparing fractions

Tell children to use the fraction wall on page 55 to help them find the answers to the questions and then write the answers in their own books.

4A, p. 56

Making fractions

Have children fold and cut a strip to make $\frac{1}{3}$s. Tell them to fold and cut each $\frac{1}{3}$ in half and see what they have made. What fraction is each of the pieces now? How many $\frac{1}{6}$ths are there for each $\frac{1}{3}$? (2) Tell them to fold and cut a strip into $\frac{1}{2}$. Now what other fractions can they make by folding and cutting each of the $\frac{1}{2}$s? (They can make $\frac{1}{4}$s, $\frac{1}{6}$ths, $\frac{1}{8}$ths etc.).

paper strips

OBJECTIVE 11.4

Fraction stories

Tell the children about a basketball team that had five players. One night just before their game, two of the players got sick. Ask the children what fraction of the team were sick ($\frac{2}{5}$) and how they worked this out (two out of 5 equal parts). Ask a child to write the fraction on the board. Now ask how many of the team were well enough to play ($\frac{3}{5}$) and how we can work this out. (There are 3 out of 5 equal parts remaining.) Write $\frac{3}{5}$ on the board.

Modelling fractions

Show the children some of the examples of fractions you have brought in and for each one ask the children what fraction is missing and what fraction is remaining. Ask them to discuss, in groups, other examples where they might see parts of wholes.

an egg carton with some eggs missing or other examples of fractions

Fractions around the room

Ask the children to look around the room or think of something they know about like the fractions they have just been talking about, e.g. $\frac{4}{5}$ of our windows are open, $\frac{4}{6}$ of my family are male, $\frac{1}{10}$ of our pigs are black etc. Ask the children to draw about their fraction in their books and to write a sentence explaining their picture.

Recognising fractions in groups

Discuss the first example and invite a child to explain what they have to do. Tell the children to write and draw their answers in their own books.

4A, p. 57

Making up to 1

Using the fraction wall and fraction pieces, the children work out questions like these: if I have $\frac{2}{5}$, how many more fifths do I need to make 1? ($\frac{2}{5}$ + ___ = 1); if I have $\frac{3}{4}$, how many more quarters do I need to make 1? ($\frac{3}{4}$ + ___ = 1). Let them explain how they used the fraction wall to help them find their answers.

UNIT 12

MULTIPLICATION 2

This topic builds on Unit 10 Multiplication 1: multiplication tables, multiplication to solve problems, formal strategies for multiplying numbers and the link between multiplying by 10 and place value.

ENGLISH WORDS/EXPRESSIONS USED

long multiplication

MAIN IDEAS

- **the strategy of multiplying by 10 can be extended to multiples of 10**
- **when multiplying large numbers we break them into their place value components (hundreds, tens and units) and multiply each of these, in stages, to reach the final answer**
- **when multiplying numbers of 2 or more digits, we start with the units digit, to make any trading that has to be done easier**
- **multiplying by numbers with two digits involves a three step process: multiplying first by the units digit, then by the tens digit, then adding these two products together**
- **long multiplication keeps track of the series of steps involved**

OBJECTIVES

The children will:

12.1 multiply numbers by ten and twenty
12.2 extend formal multiplication to include the multiplying of 100's.
12.3 interpret and solve word stories involving multiplication
12.4 explore multiplying by numbers greater than 10
12.5 use formal setting out to multiply by numbers greater than 10.

SPECIAL METHODS AND STRATEGIES

Short multiplication:

$$\begin{array}{r} {}^{1}2\ {}^{1}3\ 4 \\ \times\ 1\ 3 \\ \hline 7\ 0\ 2 \end{array}$$

Long multiplication:

$$\begin{array}{rl} {}^{1}2\ 5 & \\ \times\ 1\ 3 & \\ \hline 7\ 5 & 3 \times 25 \\ 2\ 5\ 0 & 10 \times 25 \\ 3\ 2\ 5 & 13 \times 25 \end{array}$$

OBJECTIVE 12.1

How do you multiply by 20?

Tell a story about the chef in a big hotel who needed 20 kilograms of fruit each day to feed the guests. Ask them to work out how many kilograms of fruit were needed every week. Watch what methods the children use to work out their answers. Choose several to tell the class how they did it. Accept, and praise, all methods that provide a correct answer.

materials for modelling numbers

Strategies for multiplying by multiples of 10

If none of the children has used this method, show how 7 × 20 is the same as 7 × 2 × 10. (We can split each lot of 20 into two lots of 10.) So then we have 14 lots of 10. If necessary, remind the children how to multiply by 10. (Refer to Objective 10.5.) Tell another story about the lady who has travelled 8 times this year on the bus from . . . to . . . to see her sister who is sick. The return bus fare is 20 kina. How much money has she spent altogether so far? Let the children work out the answer in pairs and observe how many of them use the multiplying by 10 strategy. Choose one child to explain how they worked it out. (Eight lots of K20 is the same as 16 lots of K10 which comes to a total of K160.)

Multiplying by 20

Discuss the example at the top of the page with the children. Ask them to explain how the picture illustrates the number sentence. Children may then use their materials to help them work out answers to the other questions, but most of the class will find this unnecessary. Correct the answers together. Demonstrate some of the answers with materials as you go.

4A, p. 58

Multiplication grids

	×10	×20	×30	×40	×40
4					
6					
11					
15					

The children work out the columns vertically, e.g. all the ×10 answers, then all the x20 answers, etc. What patterns can they see in the grid?

OBJECTIVE 12.2

Multiplying numbers more than 100

Tell the children the story about Air Niugini's four new planes that can each seat 122 passengers. How many people could the planes carry altogether? First choose children to model 4 sets of 122 people and ask another child to write the number sentence about the planes on the blackboard, vertically. Choose one child to add the materials together to find out the answer and another child to record what is happening with the number sentence on the blackboard. Remember to work out the units first, then the tens and then the hundreds. Relate the answer back to the problem ie. there are 488 people in the 4 planes.

materials for modelling numbers

Using trading

Say, 'One really hot day last week, I drank three bottles of soft drink. Each bottle held 275 mL, so how much did I drink altogether?' When they multiply the units to get 15, show them how 10 of the units are traded for a ten and the remaining units are recorded as part of the answer. This ten should then be added on to the other tens to make 22 tens altogether. Twenty of these tens are traded for 2 hundreds and added to the other hundreds in the hundreds place and the remaining 2 tens are recorded as part of the answer in the tens place. Finally, 8 hundreds are recorded in the hundreds place. Make sure that the children understand how the numbers recorded on the board relate to the materials. Relate the answer back to the problem by saying, 'I drank 825 mL of soft drink that day.'

materials for modelling numbers

Modelling and recording multiplication

Ask two children to explain how each of the multiplications in Frame A were worked out. Draw the objects on the blackboard to demonstrate what they are saying. Ask the children to use the materials to model the vertical number sentences in Frame B and to write the answer to each one. Correct their work together. Demonstrate some of the answers with materials as you go.

materials for modelling numbers

4A, p. 59

Practising with a partner

Write this sentence on the board, 'There are 132 grams of peanuts in each of the 6 packets that I bought. How many grams of peanuts have I got altogether? 'In pairs children show the 6 lots of 132 grams with their materials, and then write the number sentence vertically in their books and work out the answer (792).

materials for modelling numbers

Highest product wins

In groups of four, each child throws a dice three times and writes down the numbers thrown, e.g. 3, 5, 2. When each child has done this, they each combine two of their numbers to make a two digit number, e.g. 52 and multiply this by their third number, e.g. 3. The person with the highest product wins.

OBJECTIVE 12.3

Story Problems

A truck delivered 135 bags of rice to the supermarket one morning. If each bag contained 5 kilograms of rice, how heavy was the load altogether? Write a number sentence about this on the board, vertically. Tell the children to solve the problem with their materials, and then to work it out using numbers. The children explain how they did this and how to record it using numbers. Emphasise the links between the materials and the numbers. Make up a multiplication story about 263 × 3 and repeat the above steps making sure that all children are confident with the process.

materials for modelling numbers

Multiplying by multiples of 10

Write this sentence on the board, The teacher bought 34 boxes of pencils for the school, if there were 20 pencils in each box, how many were there altogether? Ask the children to write down a number sentence about this story and work out the answer. Watch which children remember to use the short-cut of changing 34 × 20 to 34 × 2 × 10 or 68 × 10 to get the answer 680.

materials for modelling numbers

Solving word problems

Children work out and record solutions to the problems. Read each problem and remind children that some of the problems will be easy to solve if they remember to use the short way of multiplying by 20. They may need to use their materials to help them find the answer and then write the number sentence vertically. (Answers: 1. 300; 2. 240; 3. 725; 4. K9.40; 5. individual answers; 6. 240; 7. 168; and 8. individual answers)

materials for modelling numbers

4A, p. 60

Number story book

The children make up their own story problems to solve. Tell the child to draw a picture about their story and then write the number sentence about it at the bottom of the page. Choose some of these to display in the room or make them into a book.

OBJECTIVE 12.4

Multiplying by numbers more than 10

You want to work out how much 12 lots of 15 is. Ask the children to guess what the answer might be. Say that you know the answer must be over 150 because 10 lots of 15 are 150 and we want to know 12 lots of 15. Ask two children to show what 12 lots of 15 looks like. Tell one to make ten lots of 15 and the other to make 2 lots of 15. Ask how many are in the group of 2 lots of 15?'(30) then ask how many are in the other group of 10 lots of 15. (150) Record this on the board. Repeat the previous two steps using 13 lots of 22. Let the children estimate first and see how many realise that the answer must be more than 10 lots of 22 or 220. Some children may be ready to take the next step and say that we have to add on 66 (3 lots of 22) and the total is 286. Work out the exact answer using base ten materials. Make sure you work out the units first, then the tens and hundreds.

materials for modelling numbers, e.g. base ten blocks or bundles of sticks

Multiplication practice

Discuss the illustration in Frame A. Ask children to work out what the total is and choose a child to explain their answer. Tell the children to work with a partner to find the answers to all the multiplications in Frame B.

materials for modelling numbers

4A, p 61

OBJECTIVE 12.5

Introducing long multiplication

Tell the children they are going to learn a special way of writing down number sentences about multiplication that will help them find out the answer more easily. Write 24 × 13 vertically on the board. Say that we read this as 24 multiplied by 13. Ask some children to make 13 lots of 24 . Ask two more children to arrange these as 10 lots of 24 and 3 lots of 24. While they are doing this ask the others how this will make it easier to find the answer. (Multiplying by 10 is easy and multiplying 3 × 24 is a much smaller problem.) Demonstrate how long multiplication is recorded. Start with the units. Point to the 3 lots of 24 and ask someone to work this out. (72) Show where this is recorded on the board. Now point to the 10 lots of 24 and ask how much this is. (240) Show the children where to write this. Now explain that we need to add these two amounts together to find out the total. (312) Do this. Go over the completed long multiplication to remind children of each of the steps. Tell them that this setting out is called **long multiplication**.

new word

materials for modelling numbers

Practising long multiplication

Now do another long multiplication, this time working only from the board. Write 43 × 12 vertically on the board. Remind the children that we multiply by the units first. With one hand cover the 1 in 12 while you work out 2 × 43. Write the answer on the first line. Now we multiply by 10. Write this on the next line. Now add these two lines together to find the answer. (516) Repeat using 35 × 4. This time ask the children to tell you the three main steps in order: multiply by the units, multiply by the tens, then add these together to find the answer.

Long multiplication

4A , p. 62

Work through the example at the top of the page step by step with the children. Tell the children to copy down the other long multiplications and work out the answers. Correct these together, emphasising the three main steps each time.

More practice examples

The children will need plenty of practice to master this method of multiplying. Provide them with examples like these:

27	72	36	43	51
× 14	× 15	× 24	× 23	× 14
108	- - -	144	129	- - -
- - -	720	- - -	- - -	510
- - -	- - - -	- - - -	- - - -	- - - -

UNIT 13

WEIGHT

This topic builds on measuring weights using common units and introduces one kilogram (see Teacher's Resource Book 3, Unit 7)

ENGLISH WORDS/EXPRESSIONS USED

balance, exact, weight, weigh, kilograms, grams, heaviest, lightest, convert, order

MAIN IDEAS

- **the standard unit of weight is the kilogram**
- **the weight of an object or material is not always related to its size or appearance**
- **we sometimes need to measure weight more accurately than in whole kilograms**
- **one kilogram is made up of 1000 grams**
- **the weight of an object can be in kg, g or a combination of both**
- **weight is a factor that enters into everyday situations**

OBJECTIVES

The children will:

13.1 make one kilogram and use it to estimate and weigh objects

13.2 explore the need for smaller units so that more exact measurements of weight can be made

13.3 explore the relationship between a kilogram and some parts of a kilogram ($\frac{1}{4}$ and a $\frac{1}{2}$ kilogram)

13.4 make $\frac{1}{2}$ and $\frac{1}{4}$ kilogram amounts

13.5 convert measurements in kilograms to grams and vice versa

13.6 calculate the cost of sending items of various weights

SPECIAL METHODS AND STRATEGIES

As with all measurement topics, it is important for children to hold, touch, experience, and compare practical examples. Find objects for the children to hold and discuss. A one toea coin weighs 2 grams.

OBJECTIVE 13.1

Making one kilogram

Children use their **weight** and their **balance** to make their own one **kilogram weight** using objects of their choice. They can use a bag or a container to hold their objects. When all groups have finished ask them to show the others how they made their kilogram weight. Use a balance pan to check that each does **weigh** one **kilogram.** Ensure that all children are involved.

a 1 kg weight for each group, e.g. a 1 kg bag of rice, pan balances, containers to hold objects to make a 1 kg weight e.g. bags, jars, etc.

Estimating one kilogram

Ask children to estimate and then work out in groups how many cups of wet sand it will take to balance one kilogram. Note their estimates and then let them work it out using their pan balance and the kilogram weight they have made. How close were their estimates? Now have them repeat the activity using rice. Do they estimate that it will take the same number of cups? If not, why not? Give groups time to check their estimates and then discuss what they found, giving reasons for their results. Repeat the activity using cups of leaves, stones, etc.

a cup, wet sand, rice, small stones, seeds, leaves, etc. for each group

At the market

Look at each of the pictures and guess which ones might be correct. Use the same number of real objects as shown in the pictures to check if they do weigh exactly one kilogram. Which weigh more than a kilogram and which weigh less than a kilogram? Discuss reasons for slight differences in weight, e.g. the bananas in the picture are larger/smaller etc.

3 small bananas, 6 small tomatoes, 5 kaukau, and 20 peanuts or other fruit and vegetables

4A, p. 63

Wet and dry

Balance a one kilogram weight with damp sand. Then spread the damp sand onto newspaper and leave it on a table (or in the sun). The next day pour it into the pan balance. Does it still balance the one kilogram weight? Discuss.

a 1 kg weight, damp sand, newspaper, a pan balance

OBJECTIVE 13.2

More than 1 kg

Ask one child to come out and choose an object to place on the balance. Ask if it weighs more or less than one kilogram, check with a 1 kg weight and ask how much more it might be. Let the child use some of the smaller items to balance the pans and then ask how much it weighs, e.g. 1 kg and 6 shells, 1 kg and 9 nails, etc.

a pan balance, a variety of objects that weigh a little more than one kilogram, a kilogram weight, small items such as nails, stones or shells

The need for smaller units

Children to choose some of the objects from the front and use their 1 kg weight and the smaller items to find how much the large objects weigh. They record their findings as a sentence, e.g. 3 books = 1 kg + 7 shells. Discuss their findings. Talk about the inaccuracies of giving weight measurements in this way. For example, how would you know what size stones, nails, etc. to use if weights were written in this way? Children should realise that it is impossible to get an exact weight using this method.

a variety of objects that weigh a little more than one kilogram, pan balances, 1 kg weights, small items such as seeds, stones or shells

Weighing exactly

Look at each situation and discuss why it is important to have exact measurements of weight. For example, in the first picture the nurse is finding someone's weight so she can work out how much medicine to give. In the second picture the cook is measuring ingredients for a cake. Ask, 'What might happen in these situations if the measurements were not exact?' Can children give any more examples where it is important to weigh exactly? Do the same for other situations.

4A, p. 64

Introducing grams

Tell children that the units we use for measuring these smaller amounts are called grams. Pass around the weights or other items so that children can feel their weight. As a class weigh some of the large objects again but this time use the small weights with the 1 kg weight to get a more exact measure. Write these amounts on the blackboard as kilograms and grams, e.g. 1 kg 200 g

weights or trade store items for some of these smaller amounts, for example 50 g, 100 g, 200 g, and 500 g

Cooking

Find a recipe that can be made in the classroom. Use scales to measure out the ingredients exactly.

OBJECTIVE 13.3

More or less than 1 kg?

Put the trade store items on a table at the front of the room. Ask a child to come out and choose an item. They are to use their hands and the 1 kg weight to estimate if it is more or less than one kilogram. Ask other children to choose other items from the table and do the same thing. Use a pan balance and the 1 kg weight to check that the items do weigh less than a kilogram.

about 8–10 items from a trade store weighing either 250 g, 500 g, or 750 g, a 1 kg weight, a pan balance

Parts of a kg

Look at the weight labels on the trade store items. Put them in groups according to weight. If any do not have a weight on them use one of the other items to balance it and so find its weight. Talk about what the labels mean, e.g. 500 g (or $\frac{1}{2}$ kg) means it is half the weight of something weighing a kilogram. Remind the children that the units we use to measure these smaller amounts are 'grams' and there are 1000 grams in a kilogram, so 500 grams is the same as half a kilogram. It may help them to know that the ones in the base 10 materials weigh 1 gram so 1000 of them would be about one kilogram. Also, a 1 toea coin weighs 2 g. In the same way discuss 250 grams ($\frac{1}{4}$ kg) and 750 grams ($\frac{3}{4}$ kg).

about 8–10 items from a trade store weighing either 250 g, 500 g, or 750 g.These items must be in paper or plastic containers (not glass or metal).

Making 1 kg

Children work with a partner to find two items that together would weigh 1 kilogram. Pairs can tell you their choices (500 g + 500 g, 250 g + 750 g). Check by using the pan balance and the 1 kg. weight. Now ask pairs to find three items that together weigh 1 kilogram (500 g + 250 g + 250 g + 750 g).

new word

Look at the items on the page and talk about their weights. Ask questions such as, 'What is the **heaviest** item?', 'What is the **lightest** item?' etc. Tell the children to work with a partner and write down in their own books 2 things that together weigh 1 kilogram; 4 things that together weigh 1 kilogram; 3 things that together weigh 1 kilogram; 3 things that together weigh 2 kilograms; etc.

about 8–10 items from a trade store weighing either 250 g, 500 g, or 750 g, a 1 kg weight, a pan balance

4A, p. 65

What weighs less than 1 kg?

Make a list of items from a trade store that are sold in 500 g packets and 250 g packets. Are the 500 g packets half the price of the kilogram packets?

OBJECTIVE 13.4

Making $\frac{1}{2}$ kg and $\frac{1}{4}$ kg

Tell children that their task is to make their own weights for half a kilogram and a quarter of a kilogram. Provide each group with a pan balance and something to check their weights against (500 g and 250 g packets are suitable). Give groups time to do this. When most groups have finished choose some to show their weights and tell how they made them. Encourage them to talk about any strategies they may have used to make the 250 g after they had made the 500 g or vice versa.

a pan balance for each group, containers to hold objects to make a 500 g and a 250 g weight, e.g. bags, jars etc, 250 g and 500 g weights (or items weighing 250 g and 500 g) for each group

Estimating weights less than 1 kg

Ask groups of children to choose three or four objects around the room which they estimate would weigh close to 500 g. Check each using the pan balance and their $\frac{1}{2}$ kg weight. How close were their estimates? Now ask them to estimate and then weigh three objects less than 250 g, three objects between 250 g and 500 g, and three objects between 500 g and 750 g. For this last one watch how many groups realise they need to use both their weights (500 g and 250 g) together to make 750 g. How close were children's estimates?

Weights less than 1 kg

Ask questions about the weight of the items such as, 'What is the lightest item?', 'Which items weigh $\frac{1}{4}$ kg?', 'Which two items together weigh 500 g?', etc. The children write answers to the questions on the pupil page in their books.

4A, p. 66

Weight versus size

Make a collection of items which weigh $\frac{1}{4}$ kg. Compare the difference in size between some items, e.g. 250 g of sand and 250 g of feathers. What is the smallest thing you can find that weighs 250 g? What is the biggest thing? Talk about the link between weight and size.

OBJECTIVE 13.5

Writing weights in grams

1 kg and 2 kg weights

Hold up the one kilogram weight and ask the children to tell you how many grams you would need to balance it. If necessary remind them that you would need 1 000 grams and refer to the activities from previous lessons where two 500 g weights, or four 250 g weights equalled one kilogram. Now hold up two kilograms and ask children to tell you how many grams it is equivalent to. Write the children's answers on the board as:

1 kilogram = 1 000 grams
2 kilogram = 2 000 grams

Continue until you reach 10 kilograms. Ask children to copy it into their own books.

Converting to grams

Ask children to write the weight of each item on the page, in grams. Check their answers as they are working.

4A, p. 67

Converting between kilograms and grams

new word

Choose different children to help put the boxes in **order** of weight. When this has been done discuss the way the weight is written on each packet, e.g. in kilograms or in grams. If it is written in kilograms the children must write it in grams, e.g. 1 kg 200 g = 1 200 grams. If it is written in grams the children must write it as kilograms, e.g. 1 500 grams = 1 kg 500 g, or $1\frac{1}{2}$ kg. Check as children are doing this work in their own books.

a variety of empty boxes, packets, or jars some of which have their weight marked as more than 1 kilogram

Recipes

new word

Find some recipes where children can **convert** grams to fractions of a kilogram or vice versa.

What's the difference?

Use the items on the page to ask other questions such as, 'What is the difference in grams between the oranges and the apples?'

4A, p. 67

OBJECTIVE 13.6

Postal charges

Tell the children that the cost of sending a letter depends on how much it weighs. So you must buy the correct stamps to put on the letter. Provide this table for the children and discuss it.

Weight	**Cost**
Up to 50 g	25t
51g to 150 g	58t
151g to 250 g	K1.00
251g to 500 g	K2.00

A class letter

Compose a class letter about some relevant topic and choose one child to write it on paper and another child to address the envelope. While the letter is being written show the 'pretend' letters to the other children. Pass them around and decide which is the heaviest. Put them in order of weight. Weigh the letters and use the table on the board to work out how much they would cost to mail. Repeat for the finished class letter. (If you can organise it, take the class to the Post Office to buy the stamp and mail their letter.)

3 or 4 'pretend' letters of different weights, a set of scales to weigh up to 500 g (if these are not available then use the pan balance and a set of weights, or items weighing the following amounts, e.g. 50 g, 150 g, 250 g and 500 g)

Calculating weights

Look at Frame A. Tell them that they have to use the total cost given to work out how much each letter could have weighed. For example, in Frame A the total cost of 4 letters is K1.00 so each letter must have cost 25t, therefore each letter weighed under 50 g. Ask children to estimate the weight of each letter. Do the same for Frame B and Frame C.

4A, p. 68

UNIT 14

DIVISION 1

This topic builds on sharing, division number stories, the link between multiplication and division, and sharing with remainders (see Teacher's Resource Book 3, Unit 2 and Unit 9)

ENGLISH WORDS/EXPRESSIONS USED

tables, facts, multiplication, division, remainder

MAIN IDEAS

- **division covers both the related concepts of sharing and grouping**
- **multiplication and division are opposite operations**
- **some division situations involve remainders**
- **we can use division to solve many everyday problems**
- **the formal recording of division helps keep track of the series of steps involved**

OBJECTIVES

The children will:

14.1 explore the link between multiplication and division

14.2 solve division problems for 'how many groups' and 'sharing'

14.3 solve division tasks and story problems with remainders

14.4 be introduced to the formal setting out for division and solve division problems with numbers less than one hundred

14.5 divide numbers larger than one hundred

SPECIAL METHODS AND STRATEGIES

Short division

You can set these out either way.

$5\,\underline{)\,8\,{}^{3}2}$

$\;\;\;1\;6$ rem. 2

or

$\;\;\;1\;6$ rem. 2

$5\,\overline{)\,8\,{}^{3}2}$

OBJECTIVE 14.1

Dice games

To revise the 1, 2, 3, 4, 5, 6 times **tables**, in groups children throw two dice and call out the numbers that land face up. The children must write these in their books as a multiplication sum and work out the answer. To revise the 10 times tables, they throw only one dice and multiply the number thrown by 10.

two dice

Patterns of 15

Set up 15 counters like this:

* * * * *
* * * * *
* * * * *

Ask how many rows there are, how many counters are in each row and how many counters there are altogether. Discuss with the children the different facts they might write down for this array, e.g. $5 \times 3 = 15$, $3 \times 5 = 15$, $15 \div 3 = 5$, $15 \div 5 = 3$. Most children will see the multiplication **facts** quite easily but may need help to see the **division** ones.

15 counters

Patterns of 24

Ask the children to arrange the counters into a pattern and to write down all the facts for their pattern. When groups have finished one pattern ask them to make a different pattern with the 24 counters. When all groups have finished, write the number facts they recorded on the blackboard in their families,

$3 \times 8 = 24$	$8 \times 3 = 24$	$24 \div 3 = 8$	$24 \div 8 = 3$
$6 \times 4 = 24$	$4 \times 6 = 24$	$24 \div 6 = 4$	$24 \div 4 = 6$
$2 \times 12 = 24$	$12 \times 2 = 24$	$24 \div 2 = 12$	$24 \div 12 = 2$
$1 \times 24 = 24$	$24 \times 1 = 24$	$24 \div 1 = 24$	$24 \div 24 = 1$

24 counters for each group of 4 children

Related number facts

Look at the arrays shown on the page and ask the children to write down all the multiplication and division facts about each array.

4A, p. 69

Relating multiplication and division

Hold up a card and ask children to write down the related division fact, e.g. if you hold up 5×6 they must write $30 \div 6 = 5$ or $30 \div 5 = 6$

times tables cards

OBJECTIVE 14.2

Lucky dip

Let each pair have a 'lucky dip' for a card. The children must make up a word story for their number sentence which includes the answer. Share some of the stories with the class.

a card or piece of paper with a division number sentence without the answer written for each pair of children, e.g. 25 ÷ 5

Sharing eggs

Tell the following story: '36 eggs are to be shared equally between four families. How many eggs does each family get?' Pairs solve it, using materials if they wish, and record the number sentence with the answer (36 ÷ 4 = 9). Discuss. Make sure children relate it back to the story, e.g. 36 shared between four families is 9, so each family will get 9 eggs.

How many chickens?

Tell another story like, 'I have K28. Chickens cost K4 each. How many chickens can I buy with this amount of money?' Let children use any materials they want to help them solve the story and record the number sentence with the answer (28 ÷ 4 = 7). Choose a pair to read out their number sentence and show how they did it. Choose other pairs to show how they did it especially if they used a different way. All pairs should relate it back to the story, e.g. There are seven groups of K4 in K28 so I could buy 7 chickens.

Division problems

Children to write a number sentence for each problem and then work out the answer. Let them use counters if they wish. When most groups are nearly finished ask some children to read out their number sentences and answers and write them on the board. Relate each one back to the word problem.

4A.

Division stories

Have children make up some of their own division word problems. Write them on card or heavy paper and put them in a box for the class to use as a 'lucky dip'.

OBJECTIVE 14.3

Recording remainders

Tell the children a story about putting 25 oranges into bags each of which has four oranges. Ask children to write down a number sentence for this story and to then work out how many bags would be needed, using materials if they wish. Ask children to suggest what to do with the remaining orange. Discuss, e.g. just leave it out, cut it into six equal parts, etc. Record the number sentence on the blackboard reminding children how to write the **remainder**, e.g. 25 ÷ 4 = 6 rem. 1

new word

The Remainders Game

four different coloured counters for each group of 4 children, a game board numbered 10 to 50 and two sets of five cards numbered 2, 3, 4, 5, and 10 for each group of four children

10									
11	12	13	14	15	16	17	18	19	20
21	22	23	24	25	26	27	28	29	30
31	32	33	34	35	36	37	38	39	40
41	42	43	44	45	46	47	48	49	50

Each child has a counter which they place on number 10. Shuffle the cards and place them face down. In turn, the children pick up a card and divide the number they are on, by the number on the card, e.g. if they are on 10 and they pick up the card numbered 3, their answer is 3 rem. 1. They then move forward the same number of spaces as the remainder of their division (in this case, 1 space). The first to reach 50 is the winner. Provide materials if needed.

Dealing with remainders

4A, p. 71

Ask children to write down a number sentence for each of the word problems and to copy each of the number sentences into their books. Then work out an answer for each one. Tell children to decide what to do with each of the remainders in the word problems, e.g. the 1 fish might be cut into three parts, the 2 coconuts left over might be put into another smaller bag, the galip nut might be split in two, and the K4 left over might be changed into toea and then shared between the 5 people. Check.

OBJECTIVE 14.4

Setting out division formally

base ten blocks, bundled sticks or other material for modelling numbers

Pose the following problem: 'There are 84 oranges to share equally between 4 boxes. How many oranges will there be in each box?' Choose a child to write the division number sentence for this example on the blackboard, e.g. 84 ÷ 4 = . . . linking it to the word problem as they do. Tell the children that for harder calculations like these that we might not be able to work out in our heads we use the following setting out;

4)84 or 4 $\overline{)84}$

Both are equally suitable methods. Use base ten materials or bundled sticks to show 84 (8 bundles of ten and 4 single sticks). Then show the 8 bundles shared between four boxes so there are 2 in each box and the 4 single sticks shared between four boxes so there is 1 in each box. As you show each step put the answer underneath the sum so the children can see the link between the materials and the numbers.

Divisions involving trading

State a problem such as, 'A family of three wanted to share K52 equally between them. How much would each family member receive?' Ask a child to record the number sentence on the board using the formal setting out, e.g. 3)52

Then show with materials how to work out the answer, linking each stage to the formal setting out. Starting with the 5 tens, each share gets 1 ten but there are 2 tens left over. Swap these 2 tens for twenty units so that you have 22 units altogether. Now share these so that each share gets 7 units making a total of 17 with one over. The answer is K17 with K1 left over. Discuss what to do with the remainder. It can be changed into toea and shared if children decide this is the best way.

3)5 ²2
  1 7 rem. 1

base ten blocks, bundled sticks or other material for modelling numbers

Division Practice

Tell them they must copy the sums into their book using the formal setting out for all sums. Then they are to find the answer to each of the divisions using materials to help them if they wish. After most children have done the first few, choose children to explain or demonstrate how they worked out their answers.

4A, p. 72

Open-ended division

Tell children a story like this one; 'I had some tins of fish and when I shared them between five boxes, there were two tins left over. How many tins did I start with?' The children may work in pairs, using materials such as counters to make as many different answers as they can. List their answers on the blackboard. Is there a pattern?

OBJECTIVE 14.5

Tables Practice

Write tables on the blackboard and as you write each one ask children to copy it into their book and work out the answer. Only use 2, 3, 4, 5, and 6 to divide by, e.g. 25 ÷ 5 = . . . 32 ÷ 4 = . . .

Modelling division

Tell children a story like, 'Bill earned K175. He decided to share it equally among the five members of his family. How much money would each person get?' Ask a child to write the number sentence for this on the board (175 ÷ 5). Choose another child to write it in the formal way, e.g. 5)175

base ten blocks, bundles of sticks or other material for modelling numbers

Choose two children to model 175 using two different materials such as sticks, base ten blocks or place value squares. Then show with each set of materials how to work out the answer, linking each stage to the formal setting out.

Doing division

Tell the children to copy the sums into their book using the formal setting out for all sums. Then they are to find the answer to each of the divisions using materials to help them if they wish. After most children have done the first few, choose a child to explain or demonstrate how they worked out the answer to the first sum. Repeat for the other sums.

4A, p. 73

UNIT 15

CAPACITY AND VOLUME

This topic builds on measuring capacity using common units, using one litre and parts of a litre (see *Teacher's Resource Book* 3, Unit 8).

ENGLISH WORDS/EXPRESSIONS USED

capacity, test, net, litre, container, cylinder, table, cube, prism, dimensions, layer

MAIN IDEAS

- **becoming familiar with a capacity of $\frac{1}{2}$ L and how it relates to 1 L**
- **comparing capacities—can be done in several ways. e.g. pouring and counting repeat units**
- **we can use different strategies to work out how many boxes fit into a container**
- **conservation of volume—volume depends on the number of cubes or units involved, not on their arrangement**

OBJECTIVES

The children will:

15.1 use one litre containers to make litre containers and make a one litre container and check its $\frac{1}{2}$ capacity

15.2 will compare the capacity of different containers to find out if they hold more or less than one litre

15.3 solve a practical problem of estimating capacity

15.4 explore the volume of regular shapes by counting cubes

SPECIAL METHODS AND STRATEGIES

Again the children need practical or hands on experience for this unit. Use any materials which are available.

OBJECTIVE 15.1

Estimating $\frac{1}{2}$ a litre

new word

Show one of the one **litre containers** and check that children know that a **container** of this size holds one **litre.** Now ask some children to mark the **container** to show where $\frac{1}{2}$ a **litre** would be. Have the class vote on which mark is the closest. Ask children to mark their **container** where they think $\frac{1}{2}$ a **litre** would be. Let them cut their **container** so that it holds just over $\frac{1}{2}$ a **litre**. (You may want to do the cutting if the plastic requires a very sharp knife.)

many plastic containers or drink cartons which hold one litre, scissors or sharp knives, sand or water

Testing capacities

new word

In groups, children work out a way of testing whether their container holds $\frac{1}{2}$ a litre when filled to the mark. When they have worked out their **test**, let each group show the others their method. See which groups were most accurate in making their $\frac{1}{2}$ litre container. Let them correct their mark and then trim their containers so that they hold exactly half a litre.

many plastic containers or drink cartons which hold one litre, scissors or sharp knives, sand or water

How much will this hold?

Discuss the net shown on the page and ask the children to say whether they think it would make the open-top box illustrated. Have the children copy the net onto stiff cardboard, e.g tin fish carton, cut it out and paste it together. They should be as accurate as possible. When their boxes are complete ask them whether they think their box will hold one litre or half a litre. Tell them to find a way to test how much their box holds. After they have had time to discuss a method, invite several children to demonstrate their ideas. When the children have seen that the box holds one litre explain that this size, 10 cm × 10 cm × 10 cm, was the size originally chosen to be one litre. (Keep these boxes, and the containers the children made earlier, for later activities.)

stiff cardboard, cutting tools, sand or other suitable substance

4A, p. 74

What comes in $\frac{1}{2}$ L?

Ask the children to help you make a list of things which come in $\frac{1}{2}$ litre amounts. Keep this on display.

Filling Glasses

Children estimate and then find out how many normal sized drinking glasses can be filled from a $\frac{1}{2}$ litre container.

$\frac{1}{2}$ litre container as for *Estimating $\frac{1}{2}$ a litre*, drinking glasses

OBJECTIVE 15.2

Estimating and checking capacity

new word

Hold up one of the containers from your collection and ask the children whether they think it holds more or less than one litre. Write the number of the container on the board and record the children's guesses beside this. Now ask the children how they could find out if the container holds more or less than one litre. Ask several children to explain and demonstrate their method. The method should involve using a container of known **capacity** (their one litre box or $\frac{1}{2}$ litre container and some sand).

1 litre boxes and $\frac{1}{2}$ litre containers made earlier, sand, a collection of numbered containers less than, more than and equal to one litre. Choose containers of different heights and widths.

More or less than 1 L?

Working in pairs let the children take a container from your collection and test whether it holds more or less than one litre. They should record their work in their books like this.

Container	Guess	Check
14	more than 1 L	
5	less than 1 L	

After they have tested the container they should return it and choose a new one to test. After the children have tested 5 or 6 containers, call them together and choose children to read out their findings while the others check in their books to see if they had the same result. Continue until everyone's results have been crosschecked. Refer back to the prediction made earlier to see if most children were correct in their guesses.

1 litre boxes and $\frac{1}{2}$ litre containers made earlier, sand, a collection of numbered containers less than, more than and equal to one litre. Choose containers of different heights and widths.

Comparing cylinders

new word

Ask the children to watch as you make a tall **cylinder** with the sheet of heavy paper. Now unclip the **cylinder** and make a short **cylinder** from it. Ask the children to guess which of the **cylinders** holds more. Record their predictions and ask some children to explain the reasons for their guess. Decide on a method of checking the capacity of the two **cylinders**. Have them resting on a flat surface while you pour in the sand so it does not escape from the bottom. Were most of the children accurate in their predictions?

heavy paper rectangle and clips for making a strong cylinder

Comparing capacities

Ask the children to look at the items one by one and decide whether they hold more, less or exactly one litre. Have the children record their answers in their books, like this:

4A, p. 75

Less than 1 L	1 L	More than 1 L

When they have finished, ask them to add three more items to each column in their table. Choose children to read out their results.

Comparing and predicting activities

Provide sand, funnels and a variety of containers for the children to practise comparing capacities and predicting whether a container holds more or less than one litre and more or less than $\frac{1}{2}$ a litre.

Filling containers

Provide some 1 litre containers and some smaller containers, e.g. spoons, cups, matchboxes. Let the children estimate, then check how many cupfuls (or spoons, or matchboxes) of sand it takes to fill the litre container.

OBJECTIVE 15.3

Packing boxes

Show the children a cardboard box and a can of the tinned fish. Ask them how they can work out how many cans of tinned fish can be packed into the box. Leave the box and the can where everyone can see them. Encourage them to make drawings of the box and the cans to help them. When most groups have worked out a strategy, ask them to predict how many cans will fill the box, based on their strategy. Write these estimates on the board. Choose some groups to tell the class about their strategy and how they worked out their answer. Encourage them to show the drawings they made. Pack the box with cans of tinned fish to check the children's thinking and their answers. How close were their estimates? Which strategies were the most effective?

a cardboard box, a number of tins all the same size such as tinned fish which will pack several layers high into the box

Estimating capacities

new word

Children work out how many **cubes** could be packed into their one litre box. They write down their estimate. Then they actually pack their box with **cubes**, counting the number they use. They record what they have done in their books like this.

the one litre boxes made by the children earlier, several kinds of small cubes or boxes suitable for packing into the children's one litre boxes

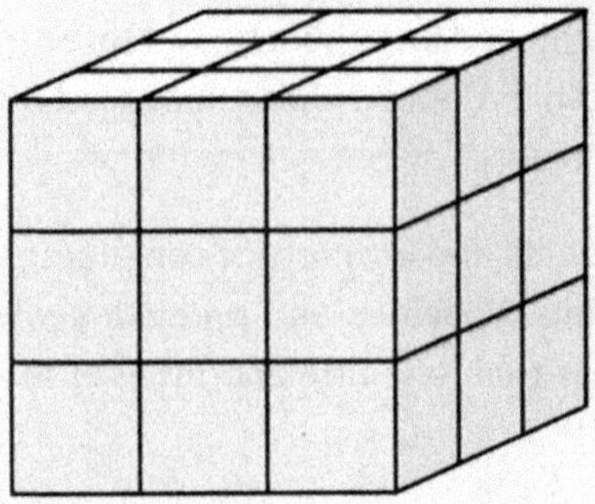

If you have more than one kind of packing cube, let the children repeat the activity using a different kind of material. Ask different groups to tell about how they worked out how many of each kind of packing cube their one litre box would hold. Has the class developed a common strategy or are there several methods being used?

Cubes in a prism

new word

Children make several 3 x 3 squares using only one colour of linking cube for each square. Now join the squares together in **layers** to make a **prism**. Ask children to write a multiplication sentence about how many cubes are in the **prism**. Repeat with other **prisms** of different **dimensions**.

OBJECTIVE 15.4

Building shapes

Ask children to build a rectangular prism that is 4 cubes high, 5 cubes long and 3 cubes wide. Ask the children to guess how many cubes there are. Write their guesses on the board. Ask them how they could work out the total without taking the prism apart to see how many are 'inside'. Discuss the children's strategies and try some of them out. If the children see the prism as being made up of layers it is a good time to practise counting in groups of 3, 4 or 5. How accurate were the guesses recorded on the board?

cubes for building shapes, e.g. linking cubes, base ten cubes or empty matchboxes will do

How many cubes?

Put the children in groups of 3 and give each group some cubes. Tell each group to make a shape that has cubes hidden inside. When they have finished, choose groups to show their model to the rest of the class who must guess how many cubes have been used in the shape.

cubes for building shapes

Hiding a cube

Tell the children to take one of their cubes and mark it in some way. Now they must make the smallest shape they can that will completely hide the marked cube inside and find out the number of cubes this takes. Give the children time to experiment and when most of them have an answer, choose groups to show and explain their solution to the rest of the class.

cubes for building shapes

Estimating volumes

Tell the children to look at the shapes and work out how many cubes would be needed to make each one. They should record their answers in their books. When they have finished choose a child to read out their answers while the others check that they agree. You may want to give out linking cubes or base ten cubes and let the children make each of the shapes to check whether they were correct.

4A, p. 76

Same volume, different shape

Ask the children to make different shapes, each one having 20 cubes. How many different ones can they make? Ask them which of the shapes is the biggest. Discuss their answers.

UNIT 16

SHAPES 1

This topic builds on recognising shapes, joining shapes, symmetry, drawing and recognising 3D shapes, and their cross sections (see *Teacher's Resource Book* 3, Units 1 and 10).

ENGLISH WORDS/EXPRESSIONS USED

copy, square, shape, circle, rectangle, triangle, diamond, trace, sides, outline, spin, flat, cone, cylinder, pyramid, face, stack, map, feature, map symbol, scale, key, cube

MAIN IDEAS

- **drawings can be copied, enlarged or reduced with some accuracy using a grid**
- **some shapes can be fitted together without leaving spaces, and some can't**
- **we give names to some common 3D objects because they have certain attributes in common, such as the number of faces, shape of sides etc.**
- **objects look different according to where we are standing when we look at them**
- **maps use symbols and scales to represent important features and their distance apart**

OBJECTIVES

The children will:

16.1 enlarge and reduce drawings using a grid

16.2 observe which and how shapes fit together

16.3 say, use and learn the names of the following regular shapes: square, circle, rectangle, triangle, and diamond

16.4 say, use and learn the names of the following solid shapes: cube, cone, cylinder, and pyramid

16.5 recognise objects based on the above solid shapes

16.6 draw solid shapes from different points of view

16.7 read, interpret and construct simple maps

OBJECTIVE 16.1

How to copy a drawing

new word

Discuss the picture on the board. Tell the children that they are going to **copy** this picture into their books using a special method. Explain how to rule up a 4 x 4 grid in their books. Start by ruling up a **square** 16 cm x 16 cm, then divide it into smaller squares, each 4 cm x 4 cm. While children are doing this, you rule a 4 x 4 grid over your drawing on the board. To copy your drawing from the board, they start at the top left hand square and copy what they see in this square onto their grid. Then they move to the next square and copy what they see in that. They work through each square on the grid until their picture is complete. Give them time to tidy up and perhaps colour in their picture. Let them discuss the results.

drawing of one of the pictures from page 77 of Pupil Book 4A on board, rulers

Make this picture bigger

Tell children that they are going to make the pictures two times as wide and two times as high. Tell them to draw two 4 x 4 grids on their paper the same way they did before, making each side two times longer than the one in the book. Tell them to copy the pictures onto their grids. Let children display their work for the others to see. Discuss in what situations it would be useful to use this technique.

rulers, paper

4A, p. 77

Practice

Provide other pictures and diagrams for the children to enlarge or reduce. Many children enjoy copying maps using this method.

OBJECTIVE 16.2

Shapes and their names

new word

Tell the children to look at the eight **shapes** at the top of the page. Read the **shape** names in the boxes. Now choose a child to name each of the shapes on the page. (The L-shape does not have a name.) Ask all the children to say the shape names together as they point to each of the shapes on the page. Discuss what shapes the children can see in the bilums at the bottom of the page. Tell the children to notice that the shapes fit together without any gaps.

4A, p. 78

Making patterns with shapes

new word

Give each child 2 shapes (select from page 78 of the Pupil Book) and tell them to **trace** and then cut out 8 (or more) of each shape. (If they fold their paper once, or twice, the job will be much quicker.) Now, tell the groups to use all their circles and six-sided shapes to make a pattern. When they have finished ask each group to report on their patterns.

a set of cardboard shapes the same size as those on page 78 of the Pupil Book, for each group of 4 children a supply of paper, such as newspaper, for cutting out shapes, scissors

Finding patterns that tessellate

Using shapes, children make a pattern with each of their shapes. They discuss the patterns made by each of the shapes in turn. Which shapes make patterns without gaps? Which ones leave gaps? Are all the patterns made with squares (triangles, rectangles, trapeziums . . .) the same or are some different from others? Give the children time to experiment with putting their shapes together in different ways. Let them choose the pattern they like best to copy into their books and colour in.

paper shapes as for *Making patterns into shapes*

Two shape tessellating patterns

If the children have not already discovered it in their free play with the shapes, introduce the idea of using two different shapes to make a pattern without gaps. What about using three or more shapes? Give the children plenty of time to freely explore the different ways the shapes can be put together. Let them paste or copy some of their favourite patterns onto paper to display in their classroom.

paper shapes as for *Making patterns into shapes*

Looking at shapes in a tangram

Allow time for children to put their five tangram pieces together to make a square. If you like you could let children work in pairs to do this. When most children have done this ask some questions which focus on each shape within the tangram. For example, 'Which piece has four equal sides? What is it called?' 'What is the name of the other piece which has four sides? What can you tell me about this shape?' etc.

5-piece tangrams, see page 217

Making different shapes with tangram pieces

Children use two of the pieces to make a triangle. (A small triangle and the trapezium, or two small triangles.) Then they use three pieces to make a triangle. (The square and two small triangles.) Then they use three pieces to make a shape like this  (This can be done in two ways: either using three small triangles; or, using the square and two small triangles.) In the same way ask them

5-piece tangrams, see page 217

to make a shape like this.  How many different ways can they do this?

Copying tangram shapes

Children use the five tangram pieces to make each of the shapes shown on the page. They work in pairs to do this if they wish. As they make each shape, they draw the position of the pieces on some paper. When most children have finished, share the results.

5-piece tangrams, see page 217

4B, p. 79

Tangram puzzles

new word

Children can make some more shapes using five tangram pieces. They should draw around them, then remove the pieces so that only the **outline** of the shape remains and let another child try to put the pieces back on to the shape.

5-piece tangrams, see page 217

Revising shape names

Divide the class into two teams. Let the children choose a partner from the opposite team. Have the children form a large circle around the shapes, each child standing beside their partner. Give each pair a stone (or similar object). Ask the children to say the names of the shapes on the ground. All the members of one team start with a throwing object. Call, 'Triangle' and the children try to toss their stone to land inside the triangle. The team scores a point for each successful throw. The children collect their stone and hand it to their partner. Call, 'Circle' and the children from the opposite team try to land their stones inside the circle. Count their score. Continue in this way until one team's score reaches 100.

one large circle, square, triangle and rectangle, drawn on the ground outside, small objects for throwing

OBJECTIVE 16.3

Spinning shapes to make spheres, cones and cylinders

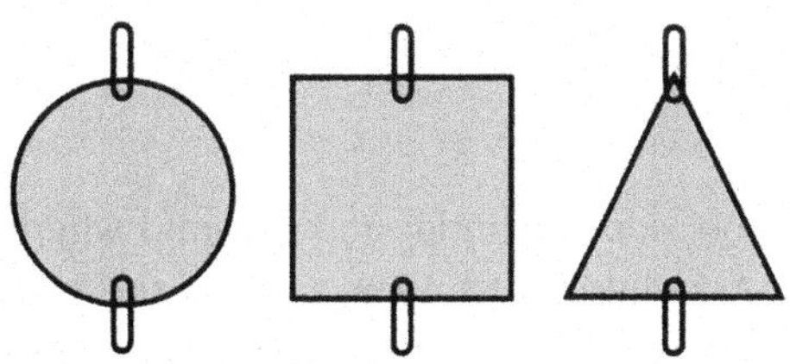

Show the children the circle and ask them to name it. Holding the elastic bands fast, wind up the circle. Tell the children to watch what shape appears when you let go the circle. Release the circle and watch it spin around. Ask the children to say what shape they saw.

a circle, isosceles triangle and rectangle cut from cardboard with elastic bands tied to either side so they can be wound up and spun

new word

If they say 'circle', remind them that a circle is a flat shape. If they say 'ball', accept this as correct. Ask them to name other objects that are shaped like a ball. Write the word 'ball'on the board and list their answers. Repeat the Introduction using the **triangle**. Introduce the word 'cone' if they do not know it and list any objects they know of that are shaped like a **cone**. Repeat the first step, using the rectangle. Introduce the word '**cylinder**' if they do not know it and list any objects they know of that are shaped like a cylinder.

Pyramids and cubes

new word

Show the children the objects you have that are shaped like a **pyramid** and a cube. Write the names of these shapes on the board and list other objects the children know that are shaped like these.

objects shaped like a pyramid and a cube

Counting faces

new word

Taking the shapes one by one, ask the children to guess how many **faces** each has and how they can work this out. Choose different children to come out and show how to count the faces on each shape. Write how many **faces** each shape has on the board beside the name of the shape.

objects shaped like a pyramid and a cube

Naming shapes

First, show examples of the shapes. Say the words, write them on the board, and have the children say the words. Choose children to name each of the objects on the page. Tell them to sketch each of the objects and write beside each sketch the name of its shape and the number of faces it has. Choose children to read out the name of each shape and the number of faces it has.

4A, p.80

Everyday shapes

Ask children to draw or bring from home any objects they have that could be added to the lists of shapes you made on the blackboard. If possible, leave these on display until the children are familiar with the shapes and can readily name them.

Practise naming shapes

Place a ball, cube, cone, cylinder and pyramid where the children can see them. Ask children to name each of the shapes and have everyone repeat the names together. Start a game of I Spy by saying, 'I Spy with my little eye something shaped like a ball.'Hold up the ball. Let the children guess until someone guesses the object you were thinking of. Remove the ball, then let this person continue the game, choosing an object based on one of the remaining shapes. Continue like this until you have used all five of the shapes. Repeat from the beginning.

examples of a ball, cube, cone, cylinder and pyramid

Comparing cubes and square prisms

Show the children the cube and ask them to name it. Show them the square prism. (Do not use the word square prism with the children.) Ask the children to compare the two shapes, e.g. how they are similar, (they both have 6 flat faces and square corners) and how they are different, (all the faces of the cube are square, the ends of the square prism are square but the sides are a rectangular shape). Ask the children to think of objects that are shaped like this. List these on the board.

a cube and a square prism

Comparing square prisms and rectangular prisms

Show the rectangular prism to the children. Compare the rectangular prism with the square prism. (Do not use the words rectangular and square prism.) Look at how they are similar, (they both have 6 flat faces, square corners and rectangular sides) and how they are different, (the ends of the square prism are square while the ends of the rectangular prism are a rectangle shape). Ask the children to think of any objects they know that are shaped like this. List these on the board.

a square prism and a rectangular prism

Comparing rectangular prisms and triangular prisms

Compare the rectangular prism and the triangular prism. (Do not use the words rectangular and square prism.) Count the faces on the triangular prism. Why is it less than the rectangular prism? (Because the end shape has only three sides, not four.)

a rectangular prism and a triangular prism

Comparing triangular prisms and hexagonal prisms

Compare the triangular prism and the hexagonal prism. (Do not use the words triangular and hexagonal prism.) Ask why the hexagonal prism has more faces than the triangular prism. (Because the end shape has six sides, not three.)

a triangular prism and a hexagonal prism. There are nets for these prisms for you to copy and make on pages 218-219.

Match the shapes

Copy the shapes in Frame A into your book. Next to each shape write the names of the objects in Frame B which have that shape. How many faces do each of the shapes in Frame A have?

4A, p. 81

Shape walk

Go for a Shape Walk in your village or local community. Ask the children to look for objects based on the 3D shapes they have learnt about in these last two lessons. They could use a record sheet like this:

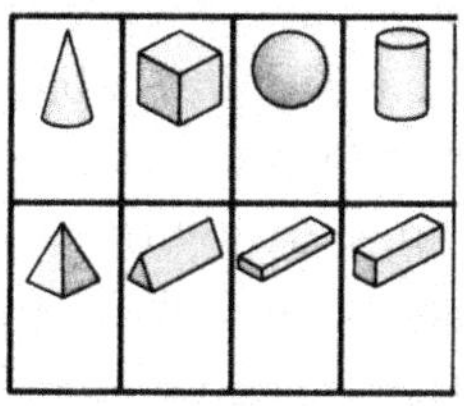

Stacking shapes

new word

Using several of your 3D shapes, let the children investigate which ones will stack up easily and which ones can be packed into a box or container. Which shapes are the best for this? Which ones do not **stack** or pack easily?

identical copies of some 3D shapes, e.g. tin fish cans

Revising the attributes of 3D shapes

Ask the children to say the names of the shapes. Tell the children to choose four of the shapes and draw them in their books. Now hold up one of the shape cards and tell the children to tick a shape they have drawn with a face shaped like this. (Shape cards are cards with plane shapes to match the faces of the 3D shapes, e.g. a circle, 2 triangles, 2 squares, 4 rectangles.) Continue like this, placing each card you have used in a pile. Any child who has ticked all four of their shapes should stand and call, 'Bingo!'. Ask them to show their shapes and check your pile of used cards to make sure there is one to match each of the child's shapes. Repeat 2 or 3 times.

models or drawings of all the 3D shapes except the ball, a set of 10 shape cards

OBJECTIVE 16.4

What would you see?

Place a chair where all the children can see it. Tell them to pretend that they are a mosquito flying over the chair: what would they see? Ask them to draw what they think the chair would look like. Choose four children to draw their pictures on the board while the rest do them in their books. When they have finished, compare their drawings. What features do all children agree can be seen from above? Are there features which some children are not sure will be visible? Place a table beside the chair and allow a student to stand on it so they have a top view of the chair. Let this student say which features of the chair can be seen from above and which drawings were most accurate.

Top views

The children choose another object in the room, without telling anyone else what they have chosen. They draw a top view of the object. When they have finished, invite children one by one to show their drawings and let the others guess what object is represented in each one. Invite children to comment on how well each picture shows the features that would be visible from above.

Different points of view

Place a large object (a chair, a stack of books) where all the children can see it clearly. Ask the class to draw what the object looks like from where they are sitting. When they have finished, choose four of the most accurate drawings, done from different points of view around the room. Shuffle these and place them on a ledge where children can see them. Ask the children to identify from where in the room each of the drawings was done. Let the owners of the drawings say whether the children's guesses are correct.

Change the point of view

Children identify each of the objects on the page. Ask children if the objects have been drawn looking from the front or the side. Tell the children to draw how each of the objects would look from a side position. When they have finished, let them compare their drawings and discuss any major differences that occur.

4A, p. 82

Point of view puzzles

Have the children draw the side view or top view of a familiar object, then draw a sketch of it. Paste these two pictures back to back. Collect all the pictures to make a set. You could make them into a book or simply have them as cards for children to play guessing games with.

OBJECTIVE 16.5

Map symbols

new word

Tell the children that you have started to draw a map on the board. Ask them to try to work out what it is a map of. Once they all know the area represented on the map, ask them to identify each of the **features** you have already drawn on the map. As each feature is identified discuss with the children the **symbol** that has been used to represent it, e.g. a green blob is a tree, a white box is a building, a blue line is a creek etc. Start writing up a key to the **symbols** beside the map you have drawn. Now ask the children what other **features** should be included on the map. Draw these in one by one, discussing what **symbol** should be used to represent each feature. You may need to add some new **symbols** to your key.

a partly completed map of a local area, e.g. school ground, village garden with scale, a long stick or strip of paper with units representing 1 metre on the map

Map scales

new word

When the map is finished, ask the children why maps are useful. (They tell us where things are in relation to each other.) Now say that maps can also tell us how big things are and how far apart they are. We can find this out from the **scale** of the map. Show the children units marked on the stick or paper strip and tell them that, on the map, this length represents 1 metre. Ask a child to come and measure something on the map using a stick or strip and then tell the class how many metres this would actually be. Now choose some other children to measure other things on the map or the distance between some of the map features and tell the class the actual distance.

a partly completed map of a local area, e.g. school ground, village garden with scale, a long stick or strip of paper with units representing 1 metre on the map

This is where I sleep

Ask the children to draw a map of the room where they sleep. They must decide what symbols they will use to represent the features of the room. They may like to work out a scale for their map. Set a time for the children to show their work to the whole class.

UNIT 17

LENGTH—SMALL

This topic builds on measuring small lengths, estimating length, perimeter, and measuring differences (*see Teacher's Resource Book* 3, Unit 4)

ENGLISH WORDS/EXPRESSIONS USED

metre, centimetre, measure, length, shorter, longer, perimeter, edge, around, compare, ruler, trundle wheel, handspan, curved line

MAIN IDEAS

- **there are a range of strategies for estimating and measuring in centimetres**
- **perimeter is the distance around an object and it can be measured in cm**
- **handspans, and other body parts, can be a convenient shortcut for roughly measuring distances**
- **measuring distances which are not straight requires different measuring strategies**
- **measurements taken in cm can be converted to m and cm and vice versa**

OBJECTIVES

The children will:

17.1 estimate and measure lengths in centimetres

17.2 measure the perimeter of different objects

17.3 use their handspan to measure objects

17.4 estimate and measure short distances, particularly those which are not straight

17.5 convert measurements written in centimetres to metres and vice versa.

OBJECTIVE 17.1

What's less than one metre?

Children look around the room and find, three things which would be less than one metre long. Ask some children to tell you what they are thinking and let them check with a one metre stick.

Making a one metre ruler

new word

Show the children how to mark **centimetres** on their stick. Each 10 cm should be labelled and the other **centimetres** just shown with a mark. Give them time to do this. Ask how many times the 10-cm strip fitted along the ruler (10). Ask how many **centimetres** this is altogether.

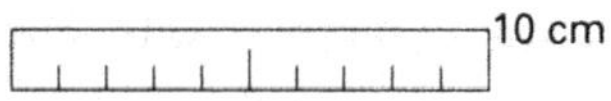

a one metre stick for each child, 10-cm strips, enough for one per child

Measuring using centimetres

new word

Without **measuring** children make a list of about ten items which they think are longer than 50 cm but less than one metre. They then use their metre ruler to check if they were correct and to write the **length** to the nearest centimetre. When most children have finished **measuring** discuss which items were correct and which were not correct.

metre rulers made in *Making a one metre ruler*

How long are these?

Children use the 10 cm line at the bottom of the page to help them estimate how long each of the objects are and then measure them using their 10-cm strip. How close were their estimates? Are any of the objects longer than 10 cm?

4A, p. 83

10cm strips

Less than 50 cm

new word

Children use the metre rulers to find items **shorter** than 50 cm. They should estimate first and then measure to see how close their estimates were.

Make one metre

Ask the children to find 3 items which, when placed end to end, measure one metre.

Estimating lengths

Put the collection of objects into the bag or box without the children seeing. Choose one child to come out and to reach into the bag or box without looking and grasp one of the objects. Let the child use both hands if they wish to explore the object and to say how long they think it is. When they have estimated the length ask them to take the object out and hold it up so everyone can see it. Ask the other children if they think the estimate is correct and then have the child use a ruler to measure it. Choose more children to select items from the bag/box until all the objects are gone.

a collection of objects between 2 cm and 20 cm in length in a bag or a box

Estimating strategies

new word

Ask pairs of children to make a list of some objects which they estimate are longer than 10 cm but shorter than 20 cm. Choose some pairs to read their list and say why they have selected those objects. Encourage children to discuss the strategies they use when they are estimating, e.g. comparing with something they know the length of. Let pairs use their metre rulers to measure the listed objects to the nearest centimetre. Were any of their objects shorter than 10 cm or **longer** than 20 cm?

a one metre stick marked in centimetres for each pair of children

Measuring in centimetres

Ask children to estimate how long the first piece of string is. Then ask them to use the scale and their string to measure it. Ask how they did it. Accept any suitable responses but emphasise that they can use the ruler marked on the page to get an accurate measurement. Have available lengths of string for children to place against the drawing of a piece of string and then to straighten out to measure against the ruler. They estimate the length of the other lines on the page and then use string and the ruler at the bottom of the page to measure them.

a 15 cm length of string for each child

4A, p. 84

Draw one metre

Let children use their 10-cm strip to draw a line which is one metre long on a page of their workbook. (The lines do not have to be straight.) Discuss how they might be able to fit a line this length on the page. After they have done the one metre line they could try to do a two metre line in a different colour on the same page.

10-cm strips

OBJECTIVE 17.2

Perimeter—the distance around

new word

Ask the children to find some things in the room the same length as their 50 cm piece of string or vine. Write their answers on the board. Now hold up the book and ask the children if they think their string will go **around** the **edge** of the book. Tell the children that we

a 50 cm piece of string or vine for each child, a collection of 4 items (books, boxes, cylindrical containers), 2 with a perimeter more than 50 cm and 2 with a perimeter less than 50 cm

call this distance all the way around, the **perimeter** of the book. Choose one child to check whether the perimeter of the book is more or less than 50 cm. Repeat with the other three items in your collection.

Strategies for measuring perimeter

new word

Ask the children to guess how many centimetres the perimeter of their table is. Record their guesses on the board. Tell the children to work out with their partner how they can measure this and let them do it. **Compare** their results with the estimates you wrote on the board. Choose several children to describe how they worked out their answer. Repeat using their Pupil Books.

rulers marked in centimetres

Measuring perimeter

Children work out the perimeter of these shapes. They should discuss with a partner a way of measuring the shapes and then measure them. Tell them to list the letters of the shapes in their books and beside each one, write its perimeter in centimetres. Record their results on the board and discuss the methods they used to find them.

4A, p. 85

Measuring perimeter in metres

new word

Using metre rulers, **trundle wheels** or string, let the children measure some larger perimeters such as the classroom, the school building, or the playground.

Same area, different perimeter

Have the children working in groups of four. Give each group four Pupil Books. (You may have to borrow some more from another grade. It does not matter if they are not Grade 4 books.) Tell the children to lay the books flat on the table side by side and measure the perimeter of all four books together. Check their answer. Now ask if there is a different way of placing the books so that the perimeter will be different. There are many ways to do this. Here are some.

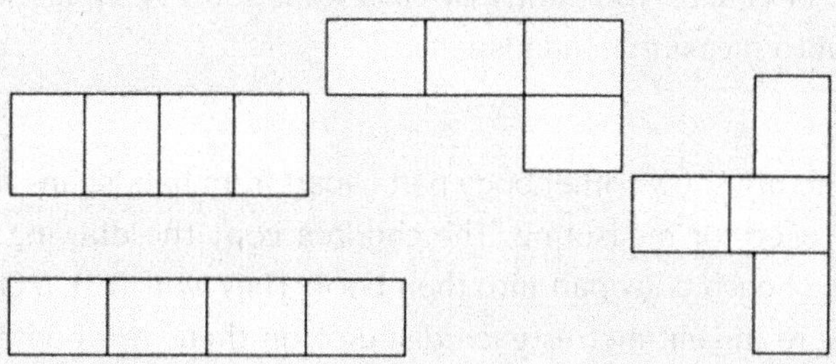

Work out which arrangement has the largest perimeter and which has the smallest.

4 pupil books for each group

OBJECTIVE 17.3

How big is my handspan?

Tell the children how, in early times, before **rulers** and centimetres were used all over the world, people sometimes measured things using their hands. Explain that a **handspan** is how far their hand can reach from the furthest part of their thumb to the furthest part of their little finger. Tell them to practise making a **handspan** and then estimate how many centimetres their **handspan** is. Let them check their estimate using a ruler marked in centimetres. Ask them if they had any difficulties such as not finishing on an exact centimetre.

new word

rulers

Measuring using handspans

Ask them to estimate how many handspans it would take to measure the length of the table. Write some of the estimates on the board. Get them to check to see how accurate they were. Now ask them to work out with their partner how many centimetres long their table is. Choose several children to tell you their answer and how they calculated it. Did everyone get the same answer? Now let the children check their calculation using a ruler.

rulers

Comparing handspan measurements

Let the children work in pairs to choose 5 or 6 objects in the room to measure in handspans. Tell them to write a list of the names of the objects and their length in handspans. When they have finished, ask a number of children to name one object they measured and its length in handspans. Make a list of at least 10 different objects that were measured on the board. Let other children compare their results with those on the board to see how close they were. Why are there differences?

What can we measure in handspans?

Children look at the list to see what things the class chose in *Comparing handspan measurements* to measure in handspans. Did anyone choose to measure the length of the room? (Too long) Or a piece of chalk? (Too short) Discuss what sorts of things would be useful to measure in handspans.

Body parts used for measuring

The picture shows other body parts apart from handspans that have been used for measuring. The children copy the drawing and the name of each body part into their book. They work with a partner to measure in centimetres each distance on their own body. Discuss as a class what each of the body parts might have been used to measure.

4A, p. 86

Research

Ask the children to find out from some adults, what words they use for describing how long, how wide, or how high things are. They could also find out how adults used to compare the lengths of two objects traditionally. Make a time for the children to report back to the whole class.Use the vernacular or lingua franca for this activity.

OBJECTIVE 17.4

Estimating and measuring perimeters

Ask the children to estimate the distance around the leg of a table or a chair. Have them write their estimate down. (Check that all children use centimetres for this.) Now discuss how you might find out how many centimetres this actually is. Children should realise that they have to place a flexible material such as string or a paper strip around the leg and then open it out and measure it against a ruler. Choose a child to do this and compare the measurement with the children's estimates.

string, wool or paper strips, a metre stick marked in centimetres

Estimating and measuring curved lines

Draw a curved line on the board, approximately 50 cm long. Ask children to estimate how long the line is and to write their estimate down. Again, discuss how you might measure its length, and choose a child to do this. How close were the estimates?

string, wool or paper strips, a metre stick marked in centimetres

Estimating and measuring in centimetres

Children estimate the length of each of the lines and write the estimates in their own books. Then they can choose string, wool, paper strips or rulers to help them measure the lines. They should write the actual measurement of each line in their books as they do it. When most children have finished, check their measurements. Whose estimates were close to the actual measurement?

string, wool or paper strips, metre sticks marked in centimetres

4B, p. 87

Challenges

Ask children to make a curved line in their own books which is exactly 30 cm long. (They could cut a piece of string 30 cm long and curve it into a shape on their page drawing a line alongside it, or they could draw a curved line and then compare it against a 30 cm piece of string adding to its length if needed.) Work out some ways of finding the distance around the edge of each of the coins.

PNG coins, string

OBJECTIVE 17.5

How far can you toss?

Take the children outside and make a starting mark on the ground. Let children take turns to stand on the starting mark and holding the bean bag with both hands at chest height throw it in front of them. They must use both hands together to throw it. Measure the distance each child throws the bean bag in centimetres, e.g. 135 centimetres.

a bean bag— a cloth bag filled with sand or rice, metre sticks marked in centimetres

Writing cm as m and cm

Record the throws of several children on the board, in centimetres. Ask how much further than one metre they each threw the bean bag. As you do, show the children how the distance can be written both as centimetres, and as metres and centimetres. For example, if the distance was 128 centimetres this can also be written as 1 metre and 28 centimetres. Ask each child to write their distance both ways. Check.

a bean bag—a cloth bag filled with sand or rice, metre sticks marked in centimetres

Converting to centimetres

Look at the long jump picture. Ask a child to read out the South Pacific Games record for this (7.49 m). Now ask another child to convert this to centimetres (749 cm). Ask that child to explain how they did this. Repeat this procedure for the other records on the page, each time asking children to explain how they converted metres and centimetres to centimetres. As children convert the records to centimetres write each one on the board.

4A, p. 88

Halving distances

Children in groups work out half of each of these centimetre distances that are written on the board. When they have done this they can change the centimetres back to metres and centimetres. Check each group's work.

How far can you jump?

Children go outside and measure out the distance for half of the long jump record. The children take turns to see if anyone can jump this far. Do the same for half the distance of the triple jump.

Measuring in the classroom

Measure the length and width of the classroom in centimetres and then change it to metres and centimetres. Choose other things in the room, more then 1 m, for the children to measure and record as both centimetres and metres and centimetres.

Partner activity

Do this activity with another child. One child uses metre sticks and 10 centimetre rulers to make a line. The other child has to calculate how long the line is in centimetres. Swap roles and repeat the activity.

metre sticks, 10 cm rulers

UNIT 18

AREA

This topic builds on finding areas of regular shapes, using area to measure land, and the square metre (see *Teacher's Resource Book* 3, Unit 9).

ENGLISH WORDS/EXPRESSIONS USED

square metre, area, compare, cover, measurements, square centimetre, grid, length, width, square units

MAIN IDEAS

- **it is useful to have a mental image of a square metre for estimating activities**
- **square metres are suitable when estimating and measuring the area of larger shapes and spaces**
- **square centimetres are used to measure smaller areas**
- **grids can be used to help us draw and calculate areas**
- **different shapes can have the same area**

OBJECTIVES

The children will:

18.1 use various strategies to compare the size of a square metre with other objects
18.2 use a square metre to describe the size of various shapes
18.3 use square centimetres as a unit of area
18.4 use centimetre grid paper to calculate the area of rectangles
18.5 use grid paper to calculate the area of other shapes

SPECIAL METHODS AND STRATEGIES

As with other measurement units, the key is to make this unit practical.

OBJECTIVE 18.1

Make a square

new word

Give each group four sticks. Children use the sticks to mark a square metre on the ground. (Angles must be right angles.) Discuss how they know it is a **square metre**. Check that each group has done this correctly. Using one of the **square metres** see how many children will fit inside it. They must have both feet on the ground.

four 1 metre long sticks for each group

How many thongs?

Groups discuss how they could work out the number of thongs they could cut from a square metre of rubber. Each group shares their strategy with the others and then tries it. (Some ways they could do it include: putting thongs in the square metre; having children stand in the square metre; cutting out shoe thong shapes to fill the space.) Groups should write their answer on the board. Discuss the answers.

four 1 metre long sticks for each group

How many tins?

Work out how many tops and bases of fish tins could be cut from a one metre square piece of tin. Groups first estimate and then work out how many would fit. Groups show how they worked it out. Discuss which way was most effective. How close were the estimates?

four 1 metre long sticks for each group, tins of fish

How many will fit?

Work out how many one litre drink containers fit into a one metre square box. Groups first estimate and then work out how many would fit. Let groups take turns to show how they would measure. How close were the estimates?

four 1 metre long sticks for each group, one litre drink containers or card cut the same size as the base of a one litre container

Covering shapes

new word

Children count how many small squares **cover** the first shape in Frame A. They use this information to estimate how many small squares would **cover** the second shape. Do the same for the other pairs of shapes on the page.

4B, p.91

Making book covers

Work out how many pupil book covers could be cut from a square metre of card. How many square metres of card would be needed to make the covers for the Pupil Books in your class? (Don't forget there is a front and a back cover.)

OBJECTIVE 18.2

Making a square metre

new word

Groups of four decide which materials other than metre sticks they could use to make a square metre. They make it and **compare** with other groups to check that they have made a square metre.

Measuring using square metres

new word

Groups first estimate and then use their square metre to measure some things around the school, e.g. the classroom, the basketball court, a table top, etc. They record their estimate and the actual **measurement** next to it. Discuss the strategies children used to estimate, and the strategies they used when measuring. Were some ways more accurate than others?

the square metres made in *Making a square metre.*

Area of a mat

new word

Groups discuss how to work out the **area** a mat would cover. They share their methods. Use a large mat or have the outline of a mat drawn in the dirt and let groups use their method to work out the **area** in square metres.

a large mat

Find the area in square metres

Children discuss how the scale measurements across the top and down the side have been used to help work out how many squares the first shape covers. Count the squares to find the area.

4B, p. 92

Rearranging square metres

Make a square metre using newspaper or cardboard. Cut or tear it and use all the pieces to make some different shapes. Do these shapes have an area of one square metre? Discuss.

newspaper or cardboard

Finding square metres

Find some different shaped objects which cover an area of about one square metre. Make a list of them.

OBJECTIVE 18.3

What do you know?

Draw a one metre square on the board or place four sticks each one metre long to make a square on the floor. Ask the children to tell you what they know about a metre. Ask children to tell you what they know about a square metre, e.g. each side is one metre long; you could fit at least twenty Pupil Books into it; the corners are right angles; etc. Write all reasonable responses.

How big is the room?

Children in groups estimate the area of their classroom in square metres. Are their estimates reasonable? Discuss how to measure the room and then do it.

metre sticks, paper cut into metre squares

Square centimetres

Hold up a Pupil Book and have children estimate its area. Accept reasonable responses such as fractions of a square metre, and discuss the difficulty of working these out. Explain how we use a unit called a **square centimetre** to measure areas that are smaller than a square metre. Show children a piece of card one square centimetre in area (1 cm x 1 cm). Children estimate how many of these squares would cover a Pupil Book. Choose two children to cover a book with these squares. Whose estimate was closest?

a Pupil Book, pieces of card one square centimetre in area

Using a centimetre grid

Explain how the centimetre **grid** can be used to work out the area of items. Children use their **grid** to work out the area of some small items. Collect small things like pens, leaves, etc. and put them on the page. Discuss what to do if the item does not completely cover a square.

4B, p. 93

Use a grid to find the area of your hand

Children should trace the grid from the Pupil Book into their own book or onto some paper. They use the new grid to work out the area of their hand by drawing it onto the grid and counting the squares. Use the same method to find the area of some other small items of their choice. Check as they are working. Discuss how they counted the parts of squares that were not whole squares.

4B, p.93

Finding area with a grid

Children place the grids over different objects and count the squares to find the area.

centimetre grids ruled on clear acetate or heavy see through plastic

OBJECTIVE 18.4

Drawing rectangles

new word

Children should trace the grid from the Pupil Book into their own book or onto some paper. Use the grid on the board to draw a rectangle with an area of 8 units. Choose a child to say how many **square units** there are inside the rectangle (8 square units). Ask children to draw on their grid a different rectangle which has an area of 8 units. Check. Can any other rectangles be drawn with an area of 8 units?

a grid drawn on the board

4B, p. 93

Finding area

new word

Draw a large rectangle on the grid. Children write an estimate of how many units are inside this rectangle. In groups children discuss how to work out how many units are inside the shape. Groups share their strategies. Possible ones are: counting the units, adding the number of units in each row or column together, multiplying **length** by **width**. (If anyone suggests multiplying the number of rows by the number of columns, or the **length** by the **width**, try it out as a class on some more rectangles to show how it works.) Use the most suitable strategy to work out the area of the large rectangle. Whose estimate was closest?

a grid drawn on the board

Areas of rectangles

Children use one of the strategies discussed to find the area of the rectangles drawn on the grid. Check that children understand what they are doing and that they are using an acceptable strategy to work out area. In particular, emphasise shortcuts for finding area, e.g. rows x columns.

4B, p. 94

12 square centimetres

Children use centimetre grid paper to draw as many different rectangles as possible which have an area of 12 square centimetres. Check.

centimetre grid paper

OBJECTIVE 18.5

Finding the area of 'L' shapes

With children watching draw a large 'L' shaped figure on the lines of the grid. Children in groups discuss how they could find, apart from counting, the number of square units the shape covers. Groups share their strategies. (Children should realise that the 'L' shape is two rectangles put together so to find its area they find the area of each rectangle and add them together.) Use this method to find the area of the 'L' shape. Draw a different size 'L' shape on the grid. Choose a child to show the two rectangles. Choose another child to work out how many units each rectangle covers and then add them together. Repeat for other 'L' shapes.

a large grid drawn on the board

Making shapes with 12 squares

Use the squares to make an arrangement like the following where all children can see it:

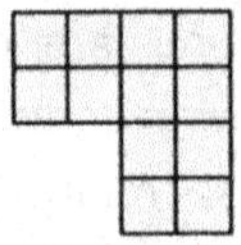

new word

Children say what the area of the shape is in **square units**. Choose a child to use the twelve squares to make another arrangement of two rectangles joined together. Does this shape also have an area of twelve **square units**? Make other shapes using 12 squares.

twelve 5 cm x 5 cm squares made out of light card

10 square shapes

In pairs children use the squares to make some shapes which have areas of 10 square units. Draw each shape.

ten 5 cm x 5 cm squares made out of light card for each pair

Calculating area

Children use a strategy of their choice to work out the area of each of the shapes drawn on the centimetre grid. Check that they are using a suitable strategy to calculate the areas of the shapes.

4B, p.95

Three rectangle shapes

Draw some shapes which are made up of three rectangles and try to work out their areas.

UNIT 19

DIVISION 2

This topic builds on the concepts of sharing and how many are in different groups, (see *Teacher's Resource Book* 3, Unit 2), and sharing with remainders, (see *Teacher's Resource Book* 3, Unit 9), and sharing story problems, (see *Teacher's Resource Book* 4, Unit 14).

ENGLISH WORDS/EXPRESSIONS USED

number sentence, multiples of ten, sequence, match, remainder, counting pattern

MAIN IDEAS

- **place value is important when dividing numbers**
- **division can involve 'sharing' or 'finding how many groups'**
- **division and multiplication are inverse operations**
- **multiplication can be used to help estimate and check the answer to a division problem**

OBJECTIVES

The children will:
19.1 divide numbers by ten
19.2 use division to solve some story problems
19.3 use multiplication to check division
19.4 learn the 8's multiplication and division tables
19.5 learn the 9's multiplication and division tables

SPECIAL METHODS AND STRATEGIES

The key to this unit is to find ways to make the calculations real. This can be done by modelling the calculations, and also by using examples from the experience of the children.

OBJECTIVE 19.1

Finding half

Pairs of children write down six numbers between 200 and 900. They discuss how to find half of each number. Talk about the ways children suggest and try them out to see which are successful. Some children will **divide** by two, others will subtract the number from itself. Discuss how to check if the method works. (Double or multiply by 2 to see if you reach your starting number.) Also discuss what happens when an odd number is halved.

Dividing by ten with materials

base 10 materials

State the problem: I have 120 peanuts. If I put 10 in each bag how many bags will I need? Pairs of children discuss how to do this. Choose a child to write the **number sentence** for the problem on the board (120 ÷ 10 = . . .). Have a child show 120 using base 10 materials. Have another child put the materials in groups of ten after **exchanging** the hundred for ten 10's. How many groups of ten are there? How many bags will be needed? Repeat for other problems where **multiples of ten** are divided by ten.

Dividing by ten

Children count to 100 by ten. Write the numbers on the board one under the other, e.g.

10

20

30 etc.

Write 'divided by 10' next to each number, e.g. 20 ÷ 10 =, and choose children to come to the board and fill in the answers. Discuss the patterns that are formed. Children should use the patterns to help them write the next ten 'sums' in the **sequence** (110 ÷ 10 = 11, 120 ÷ 10 = 12, etc.).

Investigating division by ten

base 10 materials

Children use materials to investigate if all numbers which end in 0 'behave' in the same way when divided by 10. Pairs of children choose five or six numbers which end in zero and use materials to check. Share the findings.

Using 10 toea coins

Tell this story, 'I have 80 toea in 10 toea coins. How many 10 toea coins do I have?' Have a child put out coin rubbings of 10t coins to show the 80 toea and ask another child to say how many 10t coins there are (8).

Record this on the board (80 ÷ 10 = 8) and relate it back to the story. Repeat for other coin amounts that are multiples of ten.

More dividing by ten

Children copy the sums in Frame A into their books and write the answers using either the pattern they have discovered or materials to help them. In Frame B they find the numbers which **match** and write them as a number sentence, e.g. 560 ÷ 10 = 56. In Frame C write the number sentence and then the answer. Check children's work.

4B, p. 96

Harder divisions

Divide numbers which end in something other than zero by ten, e. g. 272 ÷ 10 = 27 rem. 2. Encourage children to look for quick ways of doing these.

OBJECTIVE 19.2

Division races

a small card for each child

new word

Children write a number between 100 and 500 on their card. Collect the cards and place them face down in two piles. Divide children into two teams. The first player in each team picks up the top card from their pile, mentally divides the number on the card by 10, and says the answer out loud, e.g. if the number is 315 they say '31 **remainder** 5'. The child who first answers correctly scores one point for their team. Play until all cards are gone.

How many rows?

materials for modelling

Tell a story like, 'On a Fokker 28, there are 60 seats with 5 seats in each row. How many rows is this?' Children work in pairs using materials of their choice to model the story (some children will prefer to use real objects while others will prefer to draw a diagram). After children have modelled it they write it as a division with the answer, e.g.

5)60
   12

Ask some pairs to share how they worked it out.

Modelling problems

Children model each problem using whatever materials they want. They write the problem as a formal division in their book and calculate the answer. Choose pairs to tell the answers and show how they modelled each of the problems.

4B, p. 97

materials for modelling

Estimating division

Children practise estimating answers to divisions using numbers under 100 and divisors of 2, 3, 4, 5, or 6. Use known multiplication number facts to help, e.g. in $6\overline{)73}$ we know that $10 \times 6 = 60$, and 73 is not much more, so the answer must be a bit more than 10. In $4\overline{)65}$ we know that $4 \times 10 = 40$ and $4 \times 20 = 80$, and that 65 is a little over halfway between these numbers, so the answer will be a little more than 15.

OBJECTIVE 19.3

Checking division

Write a division on the board for children to work out however they wish, e.g. $48 \div 3 = \ldots$ Choose a child to tell you how they worked it out. Choose more children to explain their method. Accept any suitable method that produces the correct answer (16).
Children discuss how they could check that their division is correct. Choose some pairs to tell you what they think. If no one suggests multiplication show children how to use multiplication to check their division, e.g. if $48 \div 3 = 16$, then 3×16 or 16×3 will equal 48.

Linking multiplication and division

Give children the following example, $95 \div 5 = \ldots$ They work it out and then check their answer by multiplying ($95 \div 5 = 19$ so $5 \times 19 = 95$). Repeat with other examples such as $90 \div 6$ and $72 \div 4$.

Checking by multiplying

Children use multiplication to check if the divisions in Frame A are correct. If any are not correct write the sum and the correct answer in their books. In Frame B they first copy the number sentences into their books and fill in the missing numbers in as many ways as they can.

4B, p. 98

Division problems

Children discuss the story problem (question 4) and work out as many answers as possible. Provide children with other similar story problems, e.g. 'The farmer has 60 oranges to pack in bags. He cannot decide how to do it but wants the same number of oranges in each bag. Suggest some ways he could do it'.

4B, p. 98

OBJECTIVE 19.4

The 8 times table

Children, in groups, discuss which table facts with 8 they know. Do they know 8 × 1, 2 × 8, 8 × 5?

Number sentences for the 8 times table

Put out 16 items in two rows of eight. Choose children to show four ways this can be recorded; 2 × 8 = 16, 8 × 2 = 16, 16 ÷ 2 = 8, 16 ÷ 8 = 2. Children work in groups to record similar number sentences for 8, 24, 32, 40, 48, 56, 64, 72, and 80. Is there a number that children could not write four sentences for? Discuss this.

items for counting

Counting by 8

new word

Children start at 0 and write down the counting **sequence** for eight, e.g. 0, 8, 16, 24, 32 . . . up to 80. Discuss any patterns in the numbers. Children can use the **counting pattern** to continue counting by 8 as far as they can.

Silent counting

As you clap your hands slowly children count by 8 each time they hear a clap. When you stop clapping they must say what number you have stopped on. Start from different numbers and count forwards and backwards.

Linking 4's and 8's through multiplication

Children write the 4 times tables in correct order under one another. Beside them they write the 8 times tables. Discuss any links that children notice, e.g. the answer to 3 × 8 is double the answer to 3 × 4. Check if this applies to all the pairs of tables. Encourage children to use the 4 times table facts to help them work out the 8 times table.

Division patterns

Children write the division table facts for the 8's in correct order under one another, e.g.

8 ÷ 8 = 1
16 ÷ 8 = 2
24 ÷ 8 = 3 etc.

Discuss the pattern made by the answers. Investigate some other division tables to see if the same pattern occurs.

Tables races

Children line up in two teams to compete against each other. The first child from each team steps forward to answer an 8 times table fact. The child who is first to answer correctly goes to the back of their team's line and the next child steps forward. The other child remains at the front. After three turns if a child is still at the front they may go to the back of their team's row. The first team to arrive back in the starting order is the winner.

Table 'Concentration'

Spread all the cards face down. Play a game of *Concentration* until all the cards are matched. Collect the answer cards and remove one without the children seeing. Place the other answer cards face up and children have to work out which one is missing. Repeat for other answer cards.

a set of cards for the 8's division table and a set of answer cards, e.g. [56 ÷ 8] = [7]

Multiply and Divide by 8

Children copy the multiplication and division questions into their books and do the calculations. If necessary, they can model the calculation using materials.

4B, p. 99

OBJECTIVE 19.5

The 9 times nine table

Use the same activities used for the 8 times table, but substitute the 9 times table. Substitute the activity *Linking 4's and 8's through multiplication* with the activity *Linking 3's and 9's through multiplication*.

Multiply and Divide by 9

Children copy the multiplication and division questions into their books and do the calculations. If necessary, they can model the calculation using materials.

4B, p. 100

Nines on fingers

There is a system for the nine times table which works like this. Hold up your fingers on both hands. To calculate 3 times 9, fold down the 3rd finger from the left. The answer is 27.
Does this work for other nine times table facts?

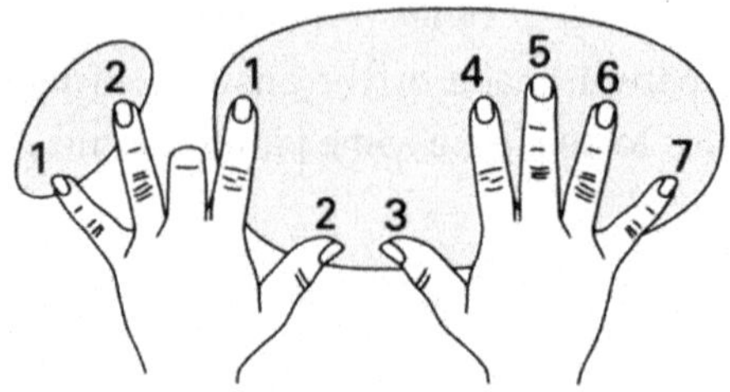

UNIT 20

PLACE VALUE TO 10 000

This topic builds on *Teacher's Resource Book* 3, Unit 3, place value to 1 000 (where children learnt to read, say, write, order and represent numbers up to 1 000 and counted large groups of objects) and *Teacher's Resource Book* 4, Unit 1, numbers in the world.

ENGLISH WORDS/EXPRESSIONS USED

thousand, place, value, represent, highest, lowest

MAIN IDEAS

- **our number system uses the same ten symbols in different orders to record any number we want**
- **our number system is based on tens**
- **zero is used as a place marker**
- **the position of a digit determines its value**
- **place value is used when ordering numbers**

OBJECTIVES

The children will:

20.1 use materials to model numbers to 10 000
20.2 use materials to make numbers to 10 000
20.3 read, write, say, and order numbers to 10 000
20.4 manipulate the value of digits within a number and recognise the value of digits in numbers to 10 000
20.5 read and write numbers and words to 10 000

SPECIAL METHODS AND STRATEGIES

- **When writing numbers in the thousands we leave a small space between the thousands and hundreds digits, e.g. 5 439.**

INTRODUCTORY ACTIVITIES

History of number

Discuss the history of our number system:
Early people recorded numbers with stone counters or by making marks on sticks or in a piece of clay. The ancient Romans used different letters for units, tens, hundreds and thousands, and they had no symbol for zero which made writing numbers very complicated. Over the years we developed easier ways to write numbers. Our number system is called the Hindu-Arabic system and uses just ten number symbols (0, 1, 2, 3, 4, 5, 6, 7, 8, 9) to record any number we want. It was first used in India in the sixth century.

The purpose of zero

Discuss its purpose: It is written as 0. When we change a number from a unit to a ten we simply add zero, e.g. 8 becomes 80. To change a ten to a hundred we add another zero, e.g. 80 becomes 800. To change a hundred to a thousand we add another zero, .. 800 becomes 8 000. This makes it very easy when writing numbers because zero can be used as a place marker. For example, in 5 086 the zero shows there are no hundreds.

Roman numerals

Have the children investigate Roman numerals, using the following guide.

I = 1, II = 2, III = 3, IV = 4, V = 5, VI = 6, VII = 7, VIII = 8, IX = 9, X = 10, XX = 20, XXX = 30, XL = 40, L = 50, LX = 60, LXX = 70, XC = 90, C = 100, D = 500, CM = 900, M = 1 000.

Choose some of these activities for children to do:

Give children some numbers to write in Roman numerals. Ask them to draw a clock-face with Roman numerals. Tell them to write out the five times table in Roman numerals. Let them write a sum in Roman numerals to give to a partner to work out the answer.

OBJECTIVE 20.1

Representing hundreds numbers

new word

Write a number on the board, e.g. 736, and ask the children to read it aloud. Say you want some children to **represent** this number using different materials. Say that **represent** means to show something in a different way. Choose three children, one to represent it on the abacus, another to represent it with place value squares and another to represent it with base 10 materials. Discuss each one. Repeat with other children for 508 and 413.

an abacus, place value squares, base 10 materials

Thousands

Write 432 on the board and ask the children to read it. Have a child represent it on the abacus. Add a thousands digit to the number on the board, e.g. 5 432. Say the new number and point out that the new digit is in the **thousands place**. Let children practise reading the number. Discuss what to do to the abacus to make this number. Have a child do it and read the new number. Do it with the other materials. Repeat for 508 and 613. Discuss how we write numbers in the thousands by leaving a small space between the thousands digit and the other digits.

an abacus, place value squares, base 10 materials

Local words for big numbers

Ask the children what words are used in their local language to talk about large numbers of things. Discuss whether these words stand for groups more than or less than a thousand. Compare and contrast the local and Hindu-Arabic numbers.

Practising reading thousands numbers

new word

Ask each child to write a number in the thousands on their card. Use these cards and some of the following activities to practise reading numbers in English. Each child reads their number to the class. The whole class reads all the numbers together. Ask which is the **highest** (biggest) number. Which is the **lowest** (smallest)? Say a number and choose a child to find that card. Six children stand and arrange themselves so that their cards are in order, then the other children read the numbers aloud to check. (Also ask which of these numbers is highest and which is lowest). One child stands, the others say the number that comes before and the number that comes after that number. Repeat with other numbers to practise the words before and after.

a card or piece of paper for each child

Using materials

Write on the board: 2 765, 4 609, 3 420, 5 088, and 3 776. Give out one type of material to each group. On 'Go' each group represents the first number. Each group scores one point if the number is represented correctly on their material and the first group to complete the number scores one extra point. Then, groups change materials and represent the second number when you say, 'Go!'. Continue in this way until all five numbers have been represented. The group with the highest score wins.

abaci, place value squares, base 10 materials

Numbers in the thousands

Explain each activity and give children time to write the answers in their books. Check their work.

4B, p. 101

Making new numbers

Use the base 10 materials and an abacus to make numbers that are 1, 10, 100, or 1 000 more or less than the numbers shown on the pupil's page. Children draw what they have done in their books.

base 10 materials, an abacus

4B, p. 101

Take six squares

Give each group of four children at least 30 place value squares (a mixture of 1 000's, 100's, 10's and 1's) in a container and some counters. In turn children take out six squares and work out what number they have. Children compare their squares. The child with the highest number wins the round and takes a counter. Repeat until one child has won five counters.

place value squares, containers, counters

OBJECTIVE 20.2

Using spear cards

new word

Use spear cards (see page 217) to make the number 743. Discuss the **value** of the 7, e.g. what it is worth? Remove the 700 card to show that its **value** is 7 hundred. Repeat for the other two digits. Now make the number 5 924. Discuss the **value** of the 5. Remove the card to show that it is 5 000. Repeat for several other four digit numbers.

spear cards

The spear card game

Groups of four children place their cards in four piles according to their value, e.g. thousands in one pile, hundreds in the next, etc. In turn children pick up a card from each pile and use them to make a number in the thousands. Children compare numbers and the child with the number closest to 2 000 wins that round and scores one point. Repeat the activity until one child has scored five points. (You could tell children to take a thousands card and then choose two, instead of three, other cards. This way children have a '0' in their numbers.)

spear cards

What numbers?

Follow the instructions on the pupil page.

4B, p. 102

Adding numbers

Have children use spear cards to make numbers that are 1, 10, 100 or 1 000 more or less than the numbers shown on the pupil page.

spear cards

4B, p. 102

OBJECTIVE 20.3

The sit down game

Each child writes a four digit number on a small piece of paper. They stand in a circle with their paper. Give some instructions such as, 'If your number is between 1 000 and 2 000, sit down', 'If your number is lower than 2 500, sit down', etc. (It is a good idea to keep a list of the instructions you give.) If a child's number fits the instruction they sit down. Continue until only one or two children are left standing. Let that child or children read their number out and check, by looking at your instructions, that they should still be standing. If so they are the winner.

small pieces of paper

Making a number line

Ask some children to help lay the cards in a line on the ground, in order, spaced as far apart as you have room. Ask each child in turn to read their number and come and place it in position along the number line. Discuss the position of each number. This is a good time to practise English words like between, higher, lower, before and after.

large number cards for the numbers 1 000, 2 000, 3 000 . . . 10 000, the numbers children used in *The sit down game*, a large space

Ordering large numbers

Follow the instructions on the pupil page.

4B, p. 103

Catalogue prices

Find some catalogues or price lists with prices in the range K1 000 to K10 000. Have children cut out and arrange the items in order of price.

catalogues, newspaper advertisements

Before and after

Children copy these numbers: 1 467, 6 809, 5 001, 7 882, 1 990, and 2 539 into their books one under the other down the centre of their page. They should write the number which comes before and the number which comes after each of the six numbers and discuss their answers.

OBJECTIVE 20.4

Make a number

Give each child a gameboard.

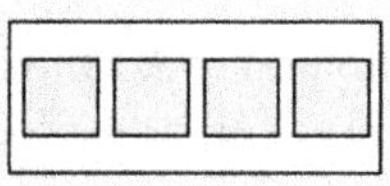

2 sets of number cards 0 to 9, a gameboard, counters

Give each group of 4 a set of cards to shuffle and spread face down in the centre. In turn, children take a card from the centre and decide where to place it on their gameboard. Once placed it must not be moved until the end of the game. This continues until each child has chosen four cards and placed them on their gameboard. Children then compare the numbers they have made. The child with the largest number wins a counter. The first child to win five counters wins the game. Discuss what strategies the children used to make the largest number they could.

Big and small numbers

Follow the instructions on the pupil page.

4B, p. 104

How many numbers can you make?

Using four different digits e.g. 5 3 7 1 children make as many different numbers as possible using all four digits each time. (There are 24 possible numbers.) They should write their numbers in order from smallest to biggest.

OBJECTIVE 20.5

Matching words and numbers

Place the number cards along a ledge. Choose a child to read out one of the numbers which you write in words on the board. Emphasise things like the word 'and', the difference between 'four' and 'forty', etc. Repeat for the other five numbers before removing the number cards. Choose different children to read the number words from the board. Hand out a number card to six children and in turn have them come to the board and match it with the words.

6 number cards
5 466 3 011 1 509
7 536 9 277 4 015

Writing numbers in

Children copy these 6 numbers into their books and write each one in words: 2 378, 4 111, 5 060, 3 602, 8 051, 6 800.

Ten more

Place the six number cards face down in a pile. A child picks up a card and writes on the board the number (in words) that is ten more. Another child says the new number in words. Repeat for the remaining cards.

the 6 number cards from *Matching words and numbers*

Writing words as numbers

Children work in pairs to do Frame A. One child reads out the words for the first number. The other child writes this as a number. Swap roles and repeat for the next number word. Continue until finished.

4B, p. 105

Match the prices

Children copy the prices from Frame B into their books and write each one in words. Children complete Frame C.

4B, p. 105

UNIT 21

SHAPES 2

This topic builds on *Teacher's Resource Book* 4, Unit 16 where children learnt the names of regular 2D shapes and dealt with tessellation and symmetry. They also learnt to name and recognise regular 3D shapes and how to draw them from different points of view.

ENGLISH WORDS/EXPRESSIONS USED

solid object,

MAIN IDEAS

- **it is important to give accurate instructions when making shapes**
- **3D objects can be drawn as 2D representations**
- **solid shapes can be modelled as skeleton shapes**
- **buildings are constructed with common 3D shapes**
- **3D shapes are made up of 2D faces of various shapes**

OBJECTIVES

The children will:

21.1 give and hear directions for making shapes
21.2 draw representations of 3D shapes
21.3 build solid shapes from sticks
21.4 discuss and draw shapes of buildings
21.5 draw shapes from particular perspectives
21.6 investigate sections of solids

SPECIAL METHODS AND STRATEGIES

- **Give children plenty of opportunity to talk about their activities so they will practise and refine their use of the English words used to describe location and the features of shapes.**
- **Children need to observe objects from different sides to see how much the appearance of a 3D object depends on our viewing point.**

OBJECTIVE 21.1

House planner

Explain to pairs of children that the four squares are to be arranged to represent the rooms in a house. They must be joined wall-to-wall not corner-to-corner, like this:

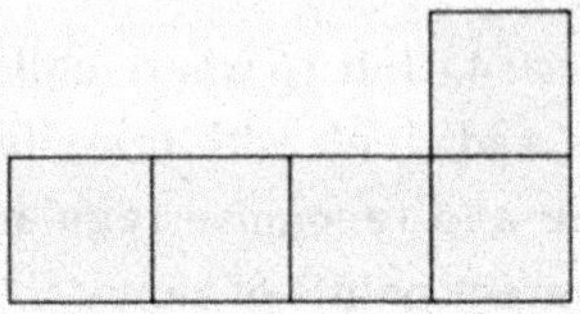

not this:

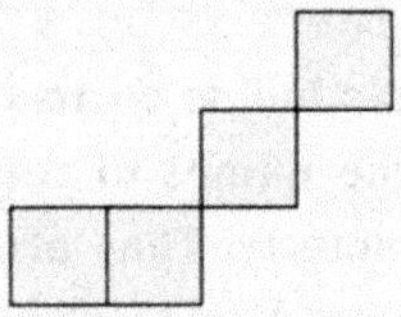

One child should give instructions in English to the other child for arranging the 'rooms', e.g. 'Put three rooms in a line and put another room to the left of the middle room.' The instruction given should describe the position of the rooms accurately and make sure the child moving the squares interprets the instructions correctly. Each child has a turn at giving instructions.

4 square shapes for each pair of children- can be cut from newspaper

Designing houses

In pairs each child draws a different design for a 4 room house without letting their partner see. One child describes their plan in English to the other so they can draw it in their book. When finished they both check to see if the house plan matches the description and what went wrong if it does not. They swap roles. Children share their house plans with the class.

Similar shapes

Children find the shapes which are similar and write the numbers for these in their books, (1 and 4, 2 and 8, 3 and 7, 5 and 6). Check.

4B, p. 106

Five triangle shapes

Children should make each shape with their paper triangles. They should use the paper triangles to make a different shape. Copy any new shapes they make into their books. Choose some children to show their drawings to the rest of the class.

5 small paper triangles for each child

4B, p. 106

Noughts and crosses

Two teams of children stand either side of the noughts and crosses grid. One member in each team gives directions in English for the first game. These two children should put their hands behind their back so they

a noughts and crosses grid large enough for children to stand in

don't point. They take turns to tell a child from their team where to stand in the grid, e.g. 'Stand in the corner square of the row closest to us, on the right hand side.' If a child can't find ways to indicate the position they want, others can suggest ways to describe the position. Play the game according to the usual rules: first team to make a line of three wins. Play as many games as you have time for.

Patterned grids

Pairs of children draw some 3 × 3 grids in their books. Each child secretly colours in some of the squares to make a pattern, e.g.

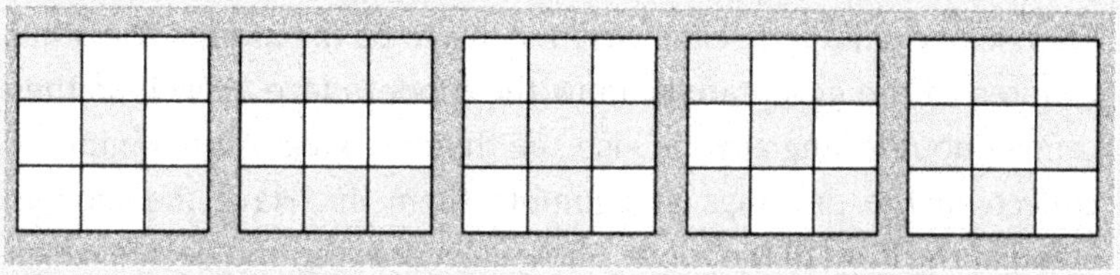

One child tells the other in English which squares to colour in on a blank grid to make the same design. Compare results and then swap roles. Play with a 4 × 4 grid or a 5 × 5 grid for a more difficult game.

OBJECTIVE 21.2

How are they different?

new word

Children look at the soap powder box, and the drawing of it on the board. They discuss the box and the drawing saying how they are different from each other, e.g. one is a **solid object** and the other is on a flat surface; the colours and the writing will look different; the sides on the drawing are parallelograms whereas the sides on the box are rectangles; you can turn the box around to look at the back; etc. Discuss how the drawing could be made to look more like the box and note the suggestions. Children look at the photo or magazine picture of the box and discuss if it looks more real than the drawing on the board. Are there things in the photo that could be added to the drawing to make it look more real? Add these to the drawing if possible so the children can see the effect.

a soap powder box, a photograph or magazine picture of the soap powder box, a drawing of the box on the board

Cartoons

Discuss if the cartoon character is a real person. How can you tell? Can you tell if the character is young or old? Can you say if it is male or female? What details in the cartoon show this? Can you tell how the character is feeling? How?

a drawing of a familiar cartoon character

What shapes can you see?

Discuss the pictures on the page. Children should name the shapes they can see in the pictures. Draw each picture in their book to show its shape as best they can.

4B, p. 107

Drawing objects

Children can draw other pictures that have the same shapes as the objects on the page.

4B, p. 107

Model faces

Choose one child from each group of five to be the model. The other children in the group are to draw the model's face as well as they can, without letting anyone else see their drawing. After 5 minutes collect all the drawings and jumble them up. Have the models stand at the front of the room. Show each drawing and decide which model is shown in each picture and what it is about the picture that shows this.

Through the magnifying glass

Children look through a magnifying glass to examine one sugar crystal and make a drawing of it. Children examine and draw other small objects such as salt crystals, a grain of sand, a grain of rice, small seeds etc.

magnifying glasses, sugar

OBJECTIVE 21.3

Making a box shape

Look at a box or carton and discuss in English the shapes of the faces. How many faces does it have? How many corners? Use sticks and joiners to make a shape like the box.

a box, small sticks, joining material such as plasticine, clay or small fruits

Comparing solid shapes and skeleton shapes

Discuss how a box and a stick shape of a box are similar and how they are different. Talk about how stick shapes emphasise the edges and corners of the shape rather than its faces.

Making shapes with sticks

Look at the shapes on the page. Have the children use their sticks and joiners to make each of the shapes. Children should close their Pupil Book and draw their stick shapes in their own books. They record how many faces and edges each shape has. Invite some children to show their model and drawing for the first shape and say how many faces and edges they counted. Repeat for the other shapes.

sticks and joiners

4B, p. 108

What shape?

Children use 12 sticks and 8 joiners to make a shape. Tell the children to describe their shape in English. Repeat for 6 sticks and 4 joiners, and 9 sticks and 6 joiners.

sticks and joiners

Make your own shape

Children use the sticks and joiners to make any shape or figure that they choose. Children keep a record of the shapes they can make by drawing them in their books and writing how many faces and edges each one has.

sticks and joiners

Shape challenge

Challenge the children to make a shape with 9 sticks and 5 joiners. Pairs of children can set challenges for each other with other numbers of sticks and joiners.

OBJECTIVE 21.4

Buildings

Look at the school building or any other interesting building in the close neighbourhood. Discuss the features of the building. What faces does it have? What shape are the faces? Can the edges of the building be seen? Children can draw the building. Discuss some of the finished drawings in terms of the features of the building that are represented.

The shape of buildings

Discuss each of the buildings on the page. How many faces are there on each of the buildings? What shape are the faces? What edges and corners can the children see?

4B, p. 109

3D shapes

Look again at the 3D shapes on page 108 and revise their features. What 3D shapes can children see on page 109 in the Pupil Book?

4B, p. 108 and 109

Making a model

In groups of three, children use the sticks and joiners to make a model of a building. They should decide what kind of building it is to be before they start: a house, a school, a shop, a tree house, a community house etc. and what features it should have. When the models are finished, children should show their model and explain the design of their building and how it would be used.

sticks and joiners

Features of buildings

Discuss why buildings have the shapes they do. Why are roofs sloped? Discuss the ways in which traditional buildings look different from modern ones. Can children suggest reasons for these differences?

OBJECTIVE 21.5

Aerial photographs

Discuss pilots who do aerial surveys and take photographs of the land below to make maps. Talk about how children could recognise the school in an aerial photograph. What would it look like? Invite some children to draw on the board what shape would be seen from the air. (It is most important that children work out the basic shape of the building but if they want to add extra details like hip lines on the roof or guttering let them do so.)

A bird's eye view

Children look at each of the buildings on the page and in turn draw what shape would be seen from the air directly above each of these buildings. The other children comment.

4B, p.110

Architect

Explain how architects need to draw a picture of a building from each side to show the builders exactly what each side looks like. Have children draw on paper what the front of your school would look like if they were standing exactly in front of the building (and could not see the side walls). They can also draw what the side of your school would look like if they were facing it directly (and could not see the front or the back). Have children take the drawings outside and check how accurate they are. What features of each **view** have been included in the drawings and what, if any, have been left out?

new word

Different views

Talk about each of the buildings on the page in terms of how many faces it has and what shape they are. What 3D shapes can the children see in the buildings? Children describe what each building would look like from the other side and say how they can work this out. They describe what each building would look like from above. Children choose one of the buildings and draw three pictures of it: one as if they were standing exactly in front of it, one as if they were standing at the side of it and one from directly above the building. When the drawings are finished, choose two children who drew the first building to show their drawings to the rest of the class and let everyone discuss them. Are they the same or different? Does everyone agree that they show what would be seen from each of the three views? Repeat for the other buildings on Pupil Book page 109.

4B, p. 110 and 109

Identifying buildings

Go for a short walk where children can see a number of different buildings. Each child chooses one building and draws one face of it, e.g. the front, the back, or one of the sides. Collect the finished drawings and choose some to show the class. Which building does each drawing represent? Identify the features of the building that can be seen in the drawing.

Looking at shapes

Have children look at each 3D shape from each side and from above. They should draw what they see each time. Do any of the shapes look the same from each perspective?

3D shapes, e.g. cone, cube, cylinder, etc.

OBJECTIVE 21.6

Cross-sections of shapes

Choose some of the kaukau shapes and discuss them with the children. Encourage the children to describe the shapes in terms of the shape and number of the faces, the number of edges and corners etc. Now ask the children to predict what shape they will see if you cut a cross-section through the shape parallel with one of the faces. Do it and check if their predictions were accurate. Repeat for other faces. Now try cutting some sections at angles to the faces to see what new shapes appear.

large pieces of kaukau cut into regular 3D shapes, a knife

Finding cross-sections

On the board draw six 2D shapes that are possible cross-sections of your pieces of kaukau. The children copy them into their books and then beside each one draw the 3D kaukau shape they think could be cut to produce a matching cross-section. Taking the shapes one by one, ask children for their predictions and then choose a child to cut each kaukau shape to see if they actually find the cross-sections they expected.

Making cross-sections

Give children time to experiment with making 3D shapes with the sand or clay and then cutting through their shapes to see what cross-sections they can find.

wet sand, clay or other suitable material

UNIT 22

TIME CALCULATIONS

This topic builds on *Teacher's Resource Book* 4, Unit 3, Time, where children learnt to read, say, and display time to one minute using both analogue and digital clocks and to use minutes and seconds to describe and estimate the duration of short events

ENGLISH WORDS/EXPRESSIONS USED

timetable, earlier, later

MAIN IDEAS

- **a calendar divides the year into months and days**
- **different cultures mark time in different ways**
- **calendars help us to keep track of time**
- **calendars are useful when working out time intervals and other time calculations**
- **timetables provide important information to help us plan our time**
- **time lines show the passage of time**

OBJECTIVES

The children will:

22.1 interpret a calendar and locate particular days, months and seasons

22.2 examine how time differences are marked in traditional systems

22.3 use calendars as the basis of a calculation of time

22.4 use a calendar to help them work out time intervals in days

22.5 calculate times before and after a given time and the interval between two times

22.6 read and interpret a variety of timetables

SPECIAL METHODS AND STRATEGIES

Here is one way to work out how many days are in a month.

Thirty days has September,
April, June and November.
All the rest have thirty-one
Excepting February
Which has twenty-eight days clear,
And twenty-nine in each leap year.

OBJECTIVE 22.1

Days in months

Identify the name of the present month and work out how many days it has. You could discuss any ways you know to work out how many days are in each month. Children can name the months and the number of days in each.

Finding special days

Display the large classroom calendar. Have children read through the names of the months as you point to them. Locate today on the calendar. Each child can find their birthday and say what day of the week it falls on. They can also find what days of the week Christmas and Easter will be this year.

a large classroom calendar for the current year

Using the calendar

Children use the calendar to show various times of the year: when it will be raining, when it will be windy, when soccer is played, the best time for planting something, the best time for catching fish, harvesting pitpit, etc.

a large classroom calendar for the current year

In the year 2010

Children look at the calendar for 2010 and work out how old they will be that year. What day of the week will their birthday be on in 2010? Children use the calendar to find particular days, e.g. what day is 30th April?, what is the last day of November?, then to find 15th February and work out how many days until the end of the month, and how many days between 15th March and 15th August? Then, children answer the questions on the page in their books.

4B, p. 111

Writing dates

Write 10 dates in numbers on the blackboard, e.g. 2.7.98, 23.10.97 etc. Tell the children that we use number digits to stand for the days, months and years. Compare this with writing the date using words. Children write the dates in words, e.g. 2nd July 1998, 23rd October 1997, etc. They can use the large classroom calendar to work out the month.

Finding Dates

Give children a paper or magazine and ask them to find as many dates as they can. Let them cut out the dates and paste them in their books. Beside each date they should write the date in numbers. You could ask them to do other things with dates they find: arrange them in order, group them according to how the date is expressed, locate them on a calendar.

old newspapers, magazines or TV guides

OBJECTIVE 22.2

Traditional time systems

Invite a local person to talk about how the days were counted in traditional time systems, and particularly about how people knew when certain celebrations or festivals were due.

Time chart of the Wedau people

Children look at the time chart of the Wedau people. They can use the chart to investigate times when there is a full moon, to find when the Wedau people plant taro, to find when the wet season is, and to find how the Wedau people know the wet season is due, etc. They can answer the questions on the page, too.

4B, p. 112

Other time charts

Obtain some time charts from some other areas and compare them to that of the Wedau people.

Time stories

Find and read some stories that tell about time and how people know when to do certain things such as planting and harvesting. Discuss them.

OBJECTIVE 22.3

Guessing months

flashcards for the months of the year

Children make teams of 3 and play against another team of 3. One team takes the cards. That team secretly chooses one of the cards. The other team asks questions to try to guess which month it is, e.g. Does the month have 30 days?, Is it a rainy month? etc. If they guess wrongly the first team scores one point. If they are right they score one point. Teams take it in turns to choose a card and to ask questions. The team with the highest score when all cards are gone is the winner.

Making a community calendar

new word

Identify events or times of the year that are significant to your local community, e.g. when the rains come, when crops are planted and harvested, when festivals are held, when fish, birds or animals are more plentiful etc. Make a list on the board and emphasise the value of this knowledge and the importance of sharing it and building up a detailed understanding of events in the community. Have the children arrange the events in the order that they happen throughout the year to create a calendar for your local community. When compiling this list emphasise English words for sequencing time such as before, after, **earlier, later,** and sequence. When the list is complete, identify the month in which each event occurs. If the list is in order put a box around all the events which fall in

January, then those that are in February, etc. Label each box so that your community calendar shows both the traditional time markers and how these correlate with the calendar of months. Display the completed calendar so that children can refer to it and perhaps add to it as the months pass.

Calendar races

Divide the class into small groups and let them have races to find particular days or dates on the calendar, e.g. What day will the 10th of September be?, What date is the first Saturday in June? etc.

4A, p. 111

OBJECTIVE 22.4

Time intervals

Children look at the calendar on the page. They use it to work out how many days it is from 24th of April until 16th of May. Choose some children to answer and to explain how they worked it out. Repeat for 12th September until 23rd December. Do the same for 3 or 4 more intervals over 3, 4, 5 and 6 months.

4B, p.113

How old are you?

Children can investigate how many days old they are. Discuss with a partner how to go about this task and how to use the calendar to help. Choose children to tell everyone the strategy they have chosen. Pairs need to have a clear strategy in mind before they start to calculate their ages in days. When the calculations are complete, each child should record their result on a class chart. Who is oldest/youngest?

OBJECTIVE 22.5

One hour earlier and later

Show 5:10 on the clock. Have a child move the hands to show 1 hour later, and explain how they worked out the new time. Repeat for 2 more times. Show 1:15 on the clock and choose a child to show what time it was 1 hour before. What time is it now? What time was it an hour ago? What time will it be in one hour?

a demonstration clock

Half hour earlier and later

Show a time on the clock. Children show times half an hour before and after the time shown on the clock. They explain how they worked it out. Repeat for half an hour before and after the present time.

a demonstration clock

Quarter hour earlier and later

Show a time on the clock. Children show times a quarter of an hour before and after the time shown on the clock. They explain how they worked it out. Repeat for half an hour before and after the present time.

a demonstration clock

Reading a timetable

new word

Check that children understand how to read the **timetable** by asking questions, e.g. When does the 7:00 plane arrive in Bialla? Children in pairs can ask each other some questions about it.

4B, p. 114

Calculating time

Have children calculate answers to questions similar to these: It takes me half an hour to get to the airport, so what time should I leave home to catch the 7:00 plane? If it takes a quarter of an hour to get to the town from the airport what time will I get to town? Choose children to answer and explain how they worked it out using the demonstration clock.

a demonstration clock

4B, p. 114

What time?

Children draw a clock to show the time they would need to leave home to catch each plane, (use the information from the previous activity). Then, they write in words the time they would arrive at the town after each journey.

4B, p. 114

Time calculations

Show a time on the clock and choose a child to work out the time one and a half hours later. Have them explain how they worked it out on the demonstration clock. Repeat for other times one and a half hours before or after a given time.

a demonstration clock

Time differences

Tell a story such as, 'I set out to walk to my uncle's house at 9:20 one day and didn't get there until 10:30. How long did it take?' Children can use the clock to demonstrate their answer. Ask children to provide stories, in their vernacular or English, about after school activities and use these to calculate how long things take.

a demonstration clock

How long did it take?

Children describe each pair of pictures on the page by saying what each story is about. They work out the difference between the two times to find how long each event takes.

4B, p. 115

Reading a TV guide

Using the TV guide ask children some questions like the following: 'How long is it between the start of the 'Nine Early News' and the start of 'Here's Humphrey'?', 'What programme starts an hour after the 'National EMTV News' finishes?', 'How long after 'Home and Away' starts does 'Tok Pisin Nius' start?' etc. In pairs, children can make up some questions of their own to ask the others.

4B, p. 115

OBJECTIVE 22.6

Working out a time problem

State the following question: A football game is to be played next week. We know the football game lasts for two and a half hours. What time could the game start and what time could it finish? Make sure all children understand the task. In groups children work at the task using one of the clocks to help them if they want. Share the answers. Record the starting time and finishing time possibilities on the board. Check that the finishing times are two and a half hours after the starting times. Discuss the earliest time that the match could start and the time that it would be most likely to start, giving reasons for the choices. Discuss the latest time that the match could finish and the time that it would be most likely to finish, giving reasons for the choices. Children should write the times from the board in order in their books from the earliest to the latest, and underline the times that they think the match is most likely to start and finish.

clocks with moveable hands

Asking questions about clocks

In pairs children work out the time shown on each clock. Pairs can make up some questions to ask others, e.g. Which clock shows the time that is 3 hours after 3:15? Children should then write the time that is half an hour before the time shown on each of the clocks, and the time that is one and a half hours after the time shown on each of the clocks.

4B, p. 116

School timetables

new word

Tell the children that the diagram is called a **timetable** and explain why. Children should discuss the purpose and usefulness of the timetable. Is this the only way it could have been drawn? Children can draw any other ways that show the information as clearly.

a weekly school or class timetable on the board showing the main events the school week

Asking questions about a timetable

In pairs children make up two or three questions about the timetable. Each pair asks one of their questions and the other children have to use the timetable to answer it.

a weekly school or class timetable on the board showing the main events of the school week

Other timetables

Ask children where else they have seen timetables. What forms of transport use timetables? Why? What forms of transport do not use timetables? Why? Make a list and collect as many examples of timetables as you can.

Make your own timetable

Children make their own timetable about what they do on the weekend. Display the finished timetables and let children ask each other questions about them.

Making a personal time line

Children make a list of the important things that have happened during their life in their family. They may like to ask their parents to help with this. They put the events in a time sequence, e.g. baby brother or sister born; started elementary school; started primary school; big sister married; first trip to town, etc. They should draw a picture about each event, write a few words to tell what is happening, label each with their age and then attach them in order on to a long strip of paper. Display the time lines and discuss how they are different from each other.

small pieces of paper, long strips of paper

UNIT 23

FRACTIONS 3

This topic builds on *Teacher's Resource Book* 4, Unit 4, Fractions 1, where children learnt to read, write, make and name common fractions, explore the relationships between parts and wholes and count using halves and quarters; and *Teacher's Resource Book* 4, Unit 11, Fractions 2, where children dealt with mixed numbers, used fraction walls to compare fractions and find equivalent fractions, and identified fractions as parts of groups.

ENGLISH WORDS/EXPRESSIONS USED

represent

MAIN IDEAS

- **the same fraction can be represented in many different ways**
- **fractions can be used to label parts of groups**
- **fractions can be used as counting labels**
- **fractions have a natural order with whole numbers**
- **estimation—a quick informed guess of the result can be used as a validity check on the final result**

OBJECTIVES

The children will:

23.1 make and interpret different representations of common fractions

23.2 use fractions to find parts of a group

23.3 count in fractions and use this to solve fraction problems

23.4 revise the use of lines to represent numbers and use lines to represent fractions

23.5 estimate what fraction is represented

SPECIAL METHODS AND STRATEGIES

It is important for children to see fractions as numbers that can be treated the same way as whole numbers—represented on number lines, included in counting sequences, added and subtracted.

OBJECTIVE 23.1

Representing fractions

new word

Write $\frac{3}{4}$ on the board. Choose a child to draw a representation of it. Discuss if it **represents** $\frac{3}{4}$. Check that the children remember that represent means to show something in another way. Have children draw different ways to **represent** $\frac{3}{4}$ and discuss each one.

Showing 3/4 with squares and rectangles

In groups children fold paper squares in as many different ways as possible to show $\frac{3}{4}$. Repeat with the paper rectangles. Do any of these representations of $\frac{3}{4}$ look like any of the ones drawn in '*Representing fractions*'?

paper squares, paper rectangles

Showing 3/4 with counters

Explain to children that fractions are not only parts of a whole but can be parts of a group. Children use some counters to show $\frac{3}{4}$. Do not tell children how many to use. Discuss how they did it. Challenge them to use some more counters to show $\frac{3}{4}$ in another way. Do they realise that they can show $\frac{3}{4}$ of any multiple of 4 counters e.g. 4, 8, 12, 16, 20, etc?

counters

Representing mixed numbers

Write $1\frac{1}{2}$ on the board and choose a child to draw a representation of it. Have children suggest other ways to represent it. Continue until there are no other ways to suggest. Discuss each representation as it appears.

Odd one out

Children look at the groups of drawings, decide which is the odd one out in each group and draw the odd one out in their books. Ask children to draw a new picture beside each of the pictures in their book to show the fraction correctly.

4B, p. 117

OBJECTIVE 23.2

People fractions

Select six children, say four girls and two boys, to stand where everyone can see them. Ask some questions about the group, e.g. What fraction of the group are boys?, What fraction are wearing shoes?, etc. Choose a child to show how each fraction is written using the symbols. Repeat for different groups of children.

Halves and thirds of 12

Spread twelve beans out. In pairs children discuss how they might give half the beans to a friend. Choose some pairs to report to the class. Put the same twelve beans into one pile. In pairs children discuss how they might work out a third of the beans to give to someone. Have some pairs explain and demonstrate how to do this.

beans or counters

Finding thirds

Give each group some counters. Write $\frac{1}{3}$ on the board. Groups use counters to find $\frac{1}{3}$ of 6 and $\frac{1}{3}$ of 9. Discuss their answers. Can groups find $\frac{1}{3}$ of 8? Choose a child to explain why this is not possible. Have children write the numbers 1, 2, 3, . . . up to 15 in their books and try to make $\frac{1}{3}$ of each number with their counters. Circle the numbers that will make $\frac{1}{3}$ and cross out the numbers which don't. Compare answers and choose children to explain their results.

counters

Fractions of groups

Write these fractions on the board: $\frac{1}{2}, \frac{1}{4}, \frac{1}{5}, \frac{3}{4}, \frac{2}{3}$. In pairs one child takes a handful of counters. The other child says one of the fractions from the board, e.g. $\frac{1}{3}$. Together children work out whether it is possible to find that fraction of that particular handful of counters. They should write any that are possible, e.g. $\frac{1}{3}$ of 15, $\frac{1}{4}$ of 16, etc. in their books. When each pair has had 8–10 turns choose pairs to tell what they found and record their fraction sentences in their books, e.g. $\frac{1}{2}$ of 14 = 7.

What fraction?

Children work out what fraction of the group is circled and record this in their books, e.g. $\frac{1}{4}$ of 12. Children should use counters, if they need, to find answers for the exercises at the bottom of the page. Check when children have finished.

4B, p. 118

Drawing fractions of groups

Ask children to think about a different way than the one shown to draw $\frac{1}{4}$ of 12. Children should draw their diagram for $\frac{1}{4}$ of 12. Discuss each one. Repeat for the other diagrams by drawing a different diagram to show the same sentence.

4B, p. 118

Fraction cards

Make some fraction cards such as, $\frac{1}{4}$ of 20, $\frac{2}{3}$ of 18, etc. for children to work out. Let them use counters if they want. In pairs children could each take a card and see who has the larger answer.

fraction cards

OBJECTIVE 23.3

Clock fractions

Point to the 12 on the clock and have children count in quarter hours as you point to the 3, 6, 9, and 12, e.g. $\frac{1}{4}$, $\frac{2}{4}$, $\frac{3}{4}$, $\frac{4}{4}$. Record the sequence on the board. Do this two or three times starting at a different point each time, e.g. the 3, 6, or 9. Discuss another way to say $\frac{4}{4}$ (1).

a demonstration clock

Counting by halves

Groups cut one square in half, diagonally. Discuss what fraction each piece represents. They cut another square of paper in half the same way and put these two pieces next to the other two pieces. Children count the pieces ($\frac{1}{2}$, $\frac{2}{2}$, $\frac{3}{2}$, $\frac{4}{2}$). Discuss another name for $\frac{2}{2}$ (1), and another name for $\frac{4}{2}$ (2). Discuss what $\frac{3}{2}$ would be (1 and $\frac{1}{2}$). Groups cut the other two squares of paper in halves and count from the start: $\frac{1}{2}$, 1, 1 and $\frac{1}{2}$, 2, etc.

4 squares of paper for each group

Counting by quarters

Groups cut each half into half again to make quarters. They count by quarters in the same way that they did for halves. Discuss another way of saying $\frac{4}{4}$, $\frac{8}{4}$, etc. Also discuss what $\frac{5}{4}$, $\frac{6}{4}$, etc would be.

the half squares from *Counting by halves'*

More/less than

Children use the diagrams to help them work out the answers. They write the answers in their books and draw a picture to illustrate their answer. Check as they are working.

4B, p. 119

Fraction sequences

Children copy the number patterns into their books and fill in the missing numbers. Check their work.

4B, p. 119

More fraction sequences

Ask children to write in their books three fraction sequences of their own (like those on p. 119 of the Pupil Book). They should swap books with a partner and add the next three numbers to each of their partner's sequences.

Fraction count

In groups of four each child in turn tosses the dice and picks up that number of thirds or quarters. They start forming circles or triangles with these thirds or quarters. When all the paper pieces have gone children should count their circles or triangles. The child who has the most is the winner. Choose children to show their patterns and count their sequence out loud.

a dice for each group, at least 12 paper circles cut into quarters per group, at least 16 paper triangles cut into thirds per group

Mixed up fractions

The cards could look like this.

$\frac{1}{3}$	$\frac{2}{3}$	1	$1\frac{1}{3}$	$1\frac{2}{3}$	2	$2\frac{1}{3}$	$2\frac{2}{3}$	3	$3\frac{1}{3}$

Mix the cards up and let children put them back in order.

a set of fraction cards for each group, each card having one fraction of a counting sequence written on it

What is missing?

Place the cards along a ledge or shelf but leave out two or three of the fractions. Children have to work out which fractions are missing and write the complete sequence in their books.

fraction cards from *Mixed up fractions*

OBJECTIVE 23.4

Fraction number lines

Discuss what each number line means. Have children copy them into their books. On the first one they should mark a place halfway between the 1 and 2. What number is represented by this? (1 and $\frac{1}{2}$) Write it on the number line. In the same way children identify 1 and $\frac{1}{2}$ on the other two number lines. Discuss why it looks like it is in a different place on each number line. Repeat for the following fractions: $\frac{1}{2}$, 4 and $\frac{1}{4}$, 2 and $\frac{1}{3}$, $\frac{3}{4}$, and 3 and $\frac{2}{3}$.

3 number lines drawn on the board: 0 to 5 with the ones marked, 0 to 50 with the 10's marked, and 0 to 10 with only the 0 and the 10 marked

More number line fractions

Identify where the number 9 is on each number line. (9 cannot be marked on the first line unless it is extended.) Where possible children mark on each of their number lines the following whole numbers and fractions: 8 and $\frac{1}{2}$, 4 and $\frac{3}{4}$, 12, 25, 5 and $\frac{1}{3}$, and 48.

the 3 number lines from *Fraction number lines*

Identify the fraction

Children look at the first group of number lines. They identify what fraction the marked point on each line represents, then write each of these fractions in their books. Check as they are working.

4B, p.120

Fill in the missing numbers

Children look at the second group of number lines. They use the marked fraction to work out what the other points on each number line are. They should re-draw each line in their books putting in the numbers or fractions for each point.

4B, p. 120

Human number line

Choose ten children to give a card to. They use the cards to make a human number line. See how long they take to put themselves in order. Repeat with other groups of ten children and with other sets of numbers. This is a good time to practise English sequencing words like: before, after, between, first, higher and lower. (These cards can also be used by individuals or small groups to practise putting numbers in order.)

ten cards with different fractions and whole numbers on each, e.g. a range of whole numbers and fractions between 10 and 15

OBJECTIVE 23.5

Estimating thirds

Show the sheet of newspaper. Children are to estimate what fraction of the newspaper they can see and write it down. They discuss with another child how they could check. (They could unfold the paper, or place it against another sheet of paper from the same newspaper to compare it.) Choose children to share their ways with the rest of the class.

a sheet of newspaper folded in thirds

Estimating eighths

Place the sheet of newspaper where children can see it. Children estimate what fraction they can see and write it down. In pairs discuss how they can check their estimate, without unfolding the newspaper, what fraction they can see. Let children try out their ideas. They can use spare sheets of newspaper if they wish.

a sheet of newspaper folded in eighths, spare newspapers

Estimating sixths

Children estimate the fraction that they can see. How can they check this without unfolding the piece of paper? Let children try out their suggestions.

an A4 sheet folded into sixths, spare papers

Estimating fractions on number lines

Children look at the first mark on the top number line. They write down the fraction they think is represented by that mark. Discuss how they can check if they are correct. Children try some of the suggested ways to see if they work. (Some possible ways are: using a ruler, base 10 pieces, string or stick to measure; using fingers to make an estimate; using your eyes to judge.) Repeat for the mark on each number line. Check the estimates.

4B, p. 121

Estimating distances

Children are to estimate certain distances along the lines, e.g. make a mark $\frac{1}{2}$ way along the line or make a mark $\frac{2}{3}$ of the way along the line. After children have made their marks check them with a ruler. Discuss strategies children can use to improve their estimates.

a page of straight lines for each child

UNIT 24

PATTERNS

This topic builds on the pattern work done in *Teacher's Resource Book* 3, lessons 1.7, 3.13, 10.10 and 10.11, where children looked at number patterns, two way grid patterns and patterns made with lines.

ENGLISH WORDS/EXPRESSIONS USED

fortnight

MAIN IDEAS

- being aware of number patterns allows us to predict the next number in a sequence
- counting by different numbers forms different patterns
- we can use a visual pattern to help us remember numbers
- number patterns can be described in words and translated into other patterns such as shape or action patterns
- mathematical problems are easier to solve if we can identify and use the patterns they contain

OBJECTIVES

The children will:

24.1 count forwards and backwards, beyond 1 000 to 10 000 by 1, 2, 5, 10, and 100

24.2 use final digit patterns to help them count forwards and backwards to 10 000 by 1, 2, 5, 10, and 100, from any starting point

24.3 practise remembering numbers, particularly those with patterns

24.4 identify, continue and describe number patterns

24.5 identify patterns in physical situations and use these patterns to solve problems

24.6 learn the 7 times table

Special methods and strategies

Encourage children to represent situations involving patterns with a number sequence, diagram or picture, which will usually make the pattern more obvious and the problem easier to solve.

OBJECTIVE 24.1

Counting to 10 000

Children sit in a circle. Choose a child to start the count and continue counting around the circle by 100's, e.g. 100, 200, 300, 400, . . . etc. until a child reaches 1 000. Discuss what will come next (1 100) Write the sequence on the board, discussing the two different ways of saying numbers over one thousand, e.g. one thousand, or ten hundred; and eleven hundred, or one thousand and one hundred. Continue the count until you reach 2 000, and do the same as before, writing the next number in order on the board. Keep going until you reach 10 000.

Counting by tens

Explain to children that you are going to start at 1 500 and count by tens silently. Every time you count another ten you clap your hands so they know. They have to keep count in their heads so that when you stop clapping they know what number you have counted to. Do this slowly at first and only count about five or six numbers from the starting number (use a different one each time) until children get used to it. When they are confident make it faster and count more numbers.

Silent counting

Repeat the *Counting by tens* activity but count by 1's, 2's, 5's and 100's from any starting point beyond 1 000 to count forwards and backwards. Each time you stop clapping children should write down the last number counted. Choose different children to say the answer each time.

Which number is 10 more?

Children should write the number that is ten more than the number on the left. They could also write the number that is 100 less.

4B, p. 122

Complete the grid

Children copy this grid into their books and complete it.

4B, p.122

Calculator counting

Show the children how to count using the constant function, e.g key in 1 230 and count by 10 by pressing + 10 = = = =

a calculator

Counting money

Ask questions like, 'I have K2 346 in my bag. How much will I have if I put in another eight of these K10 notes?' Give the children time to work it out, ask for their predictions and then count together as you put eight K10 'notes' in the bag. Repeat using different bundles of K2, K5 and K10 'notes' each time.

a bag and some pieces of paper to represent K2, K5 and K10 notes

OBJECTIVE 24.2

Number sequences

Write a number on the board, e.g. 2 476 and have children copy it into their books. When you say 'Go' they write the next ten numbers counting by 5 (2 481, 2 486, 2 491, 2 496, 2 501) . Check the sequence of numbers and discuss what pattern the final digits make when we count by 5. (Make sure the children can identify which are the final digits, and that they can recognise and understand this expression in English.) Repeat for other numbers this time counting by a different number, e.g. 2 or 10, or going backwards, each time discussing the pattern made by the final digits.

How much money?

4B, p. 123

Children work with a partner to find out how much money will be in the bag if the K10 notes are put in. They write the answer in their books. Repeat for the other boxes.

Counting patterns

4B, p. 123

In pairs, children copy the counting sequences and write in the missing numbers. They then make up two counting sequences that start and finish with the numbers shown in the Frame at the bottom of the page and write these sequences in their books. Choose some pairs to read out their answers.

Final digit patterns

Jumble up each set of number cards, keeping the sets separate. Put children into three groups and give one set of cards to each group. Children share out the cards so that each child has one. Explain that each set of cards makes a counting sequence. Groups must work out together what the sequence is and then stand themselves in order, each holding a card. When groups have done this read out each sequence to make sure they are correct. Have each group explain how they worked out what their sequence was. Did looking at the final digits of their numbers help them?

3 sets of about 10 cards, each set with a different counting sequence, e.g.

5 415, 5 420, 5 425 5 430

3 762, 3 764, 3 766, 3 768

How much will be left?

Give the bag to a child and say, 'There is K5 670 in here. How much will be left if 'Tobias' gives each of you a K10 note?' Give pairs of children time to write down some of their predictions. Let 'Tobias' walk around giving out K10 notes and counting back by 10 each time to see if the predictions were correct. Repeat for other amounts.

a bag and some pieces of paper to represent K2, K5 and K10 notes

OBJECTIVE 24.3

Numbers we remember

Point out that there are many numbers that people need to remember. Ask pairs of children to work together to make a list of some of these, e.g. phone numbers, dates, distances, Post Office box numbers, number plates, prices, flight numbers, account numbers, pin numbers, Bible verses, hymn numbers and radio stations. Share the lists and make a class chart.

Strategies for remembering

Choose some children to tell you a date that is important to them. Write some of these on the board. Cover one of the dates and ask the class to remember what it was. Have the child who gives the correct answer tell how they remembered. Discuss how it is easier to remember a date if we break it into parts and look at the day, the month and the year. Is it an early or late day in the month? What is the name of the month? Is the year in the past, the future or is it the present year and so on. It also helps to notice patterns in dates such as double figures (22.6.99) or counting patterns (2.4.68). Let the children have another minute to look at the dates on the board and then cover another one of them. Are more children able to remember the date this time? Repeat for other dates.

Did you remember?

Hold up the first of the number cards for 10 seconds then remove it. The children are to write the number in their books. Have several children say what they think the number was, then show the card again to see who was correct. Hold up the next card. Give them 10 seconds to look at it and then write the number in their books. Who was correct? Repeat this with the last number card. Discuss ways that children found it easy to remember the numbers.

3 number cards:

| 23756784, |
| 2179 6998, |
| 23 88 67 09 |

Remembering numbers

Children work with a partner. One child holds up a card for 10 seconds while the other studies the numbers and then checks to see how well they remembered it. Swap over and continue.

more cards like those used in *Did you remember?*

Where have you seen these?

In pairs, children look at the numbers and discuss where they might see numbers like these. Ask children to talk about ways of remembering some of the numbers on the page. Share the ideas as a class. Pairs should then look for patterns or other features of the numbers that might make them easy to remember. Talk about how numbers have more meaning and are easier to remember if we can relate them to something we know.

4B, p. 124

Write the numbers

Children have 5 minutes to write as many of the numbers that were on the page as they can remember. After 5 minutes, they stop writing. Choose children to read out their answers. Tell them which numbers they read out correctly. Children add up how many they remembered. They open to page 124 again to check the numbers they missed out on. Invite children to say what it was about the numbers that helped them to remember them easily.

4B, p. 124

Phone numbers

Look in telephone directories or at the advertising sections in newspapers and magazines and collect phone numbers of companies and businesses. Also look at the phone numbers for emergency services like the police and ambulance. What do children notice about some of these phone numbers? What phone number would they choose, if they could, to make it easy for themselves and others to remember it. You might like to make a chart displaying some of these phone numbers.

telephone books

OBJECTIVE 24.4

Describing number patterns with shapes

Write the following pattern on the board: 1 2 2 1 2 2 1 2 2 1. Choose some children to describe the pattern. In pairs children should make a different pattern using any two numbers, e.g. 7 7 7 8 7 7 7 8 7 7 7 8. When finished have them describe their pattern. Hand out the counters or shapes. Pairs are to use them to make the same pattern, e.g.

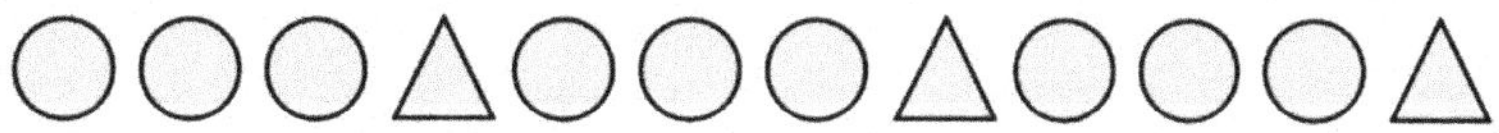

Look at the finished patterns. Choose children to make their pattern using claps and finger clicks. Invite other children to make their pattern in any other way they can think of.

two different kinds of shapes or counters

Describe your pattern

Groups of four, in pairs take turns to make up a rule for a pattern, e.g. start with an even number and keep adding 5, e.g. 34, 39, 44, 49, 54, 59. They say their rule to the other pair who write the pattern that fits the rule. The first pair check that they agree with the pattern. Swap roles and continue.

Patterns and rules

There are patterns down one side and a rule for each pattern down the other side. Children copy each pattern into their book, complete it, and write the matching rule for it alongside each pattern.

4B, p. 125

Find the rule

In pairs children think of their own rule and write down the first 8 numbers in their pattern. Choose one pair to read out their number sequence for you to write on the board. The others try to work out the 'rule' and the next two numbers in the sequence. Encourage them to work out the difference between numbers in the sequence to help them. Choose others to share their number pattern with the class. Collect the remainder and make them into a book, with solutions on the back of each page, for the children to do as an independent activity.

Calendar patterns

calendars

Have a look at the numbers on a calendar. What pattern can the children see in the numbers in each column? (+7, +7 . . .) Can they explain why this is? Now draw a box around any 4 numbers, e.g.

2	3	4	5
9	10	11	12

Add the numbers in the box diagonally like this;

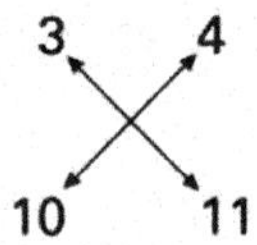

What happens? Choose any four numbers on the calendar and repeat. What happens? Can anyone explain why?

Clapping patterns

In pairs children make clapping patterns for the class to follow. Each pair presents their pattern and the others join in.

Matchstick squares

lots of matchsticks or pieces of stick all the same length

Have children make a square using one matchstick for each side. How many were used? Make an adjoining square, e.g.

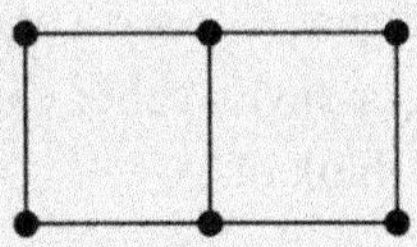

How many matchsticks were used? In pairs, children work out how many matchsticks it will take to make 3, 4, 5, etc. squares. They should record their results in their books. Discuss the results. Were there any patterns? Repeat this activity but make triangles instead of squares.

Tyre and shoe patterns

Children in pairs investigate the marks made by tyres or shoes. Do all shoes or tyres have the same tread pattern? What shapes are the patterns made of?

Shadows

Groups of children place a stick upright in the ground in a position where it receives direct sunlight. Each hour through the day they measure the length of the shadow cast by the stick and record it in their books. Discuss what happens to the shadow over a period of a day. Does the shadow follow a pattern?

long sticks and measuring sticks or tapes

OBJECTIVE 24.5

The magic bean

Discuss this problem: A magic bean plant is 2 metres tall. Each week it grows 2 more metres. How tall will it be after one week? (4 m) After two weeks? (6 m) After three weeks? (8 m) What about after 10 weeks? Children in pairs discuss how to work out this last part of the problem. They then share their solution and strategies for working it out, discussing how the pattern helped them.

Filling cans

Tell this problem: At the oil refinery, Joe started the day with 6 785 litres of petrol in the main tank. His job for the day was to fill empty 50 litre cans of petrol. Children in pairs work out how much petrol would be left in the main tank after Joe had filled one can, then after he had filled 2, 3 and 4 cans. They can also work out how many cans Joe had filled by lunch time when the meter on the main tank showed there were 6 035 litres left. Choose some pairs to show how they worked it out and how they used the pattern to help them.

Using number patterns

Children read the problems and explain in their own words what each problem is asking them to do. In pairs children solve each of the problems. They record their working out in their books, by setting out each problem in the form of a number pattern. Discuss how the patterns helped them to find the answers.

4B, p. 126

OBJECTIVE 24.6

Counting by 7's

new word

Discuss with children the number of days in one week, and the number of days in two weeks or a **fortnight**. Write these on the board one under the other and write the number of days beside them. Continue the table, listing 3, 4, 5 . . . 10 weeks. With a partner children are to work out how many days are in each of these. Write the answers on the board beside the weeks. As a class use the list to count by 7's.

Seven times table

Put children into two teams. The first person from each team tries to be first to answer questions about the number of days in weeks, e.g. how many days in 6 weeks? The winner scores a point for their team. Continue in pairs until all children have had at least one turn.

Sevens again

Explain to the children that you are going to count by 7's in your head starting from 0 and clap each time you say a number to yourself. When you stop clapping the children must tell you what number you have reached. Do this a few times stopping at different numbers each time. Repeat the activity starting from 70 and going backwards.

7 times tables 'Concentration'

Place both sets of cards face down. Children play a game of *Concentration* until all pairs of cards are matched.

a set of cards for the seven times table, e.g. 6 x 7, a set of answer cards for the seven times table, e.g. 42

Missing sevens

Using one set of answer cards, remove one card without children seeing, then place each of the other answer cards face up and ask children to say which one is missing. To do this children will need to count by 7's. Repeat for other numbers.

a set of answer cards for the 7 times table

Tables races

Children line up in two teams to compete against each other. The first child from each team steps forward to answer a 7 times table fact. The child who is first to answer correctly goes to the back of their team's line and the next child steps forward. The other child remains at the front. After three turns if a child is still at the front they may go to the back of their team's row. The first team to arrive back in the starting order is the winner.

UNIT 25

USING FRACTIONS

This topic builds on the recognition of fractions, and seeing fractions as parts of wholes.

(See *Teacher's Resource Book* 4, Unit 4 Fractions 1, Unit 11 Fractions 2 and Unit 23, Fractions 3.)

MAIN IDEAS

- **half, quarter and eighths are fractions, which show amounts or numbers and they can be added and taken away.**
- **important quantities like kilogram, hour and minute can be used in parts, and we can have $\frac{1}{2}$, $\frac{1}{4}$ and other fractions of them.**

OBJECTIVES

The children will:

25.1 make a system of representing fractions and use it to solve word problems and to add fractions.

25.2 be introduced to subtraction of fractions using fraction squares.

25.3 find fractional parts of kilograms and hours.

25.4 create fractions in different ways.

SPECIAL METHODS AND STRATEGIES

The key in this unit is using fraction squares to aid calculations with fractions.

The children should use the fraction squares for all calculations.

For the kilogram and hour, have some practical examples to show such as a sugar packet and a clock.

OBJECTIVE 25.1

Fractions of a square

heavy paper, fraction squares

Each group has five heavy paper or card 8 cm x 8 cm squares, some scissors and a ruler. Keep one set of squares for yourself to demonstrate what to do. Instructions of how to fold and cut the squares are in Pupil Book 4B pages 179–180. Tell groups to pick up two of the squares and put them to one side as they will remain whole. They take another square, use a pencil to rule a line to divide it in half and then carefully cut it in half. Put the two halves to one side. Pick up another square and after ruling it, carefully cut it into quarters. They cut the last square into eight equal pieces to make eighths.

Hold up one of your whole squares and say, 'If this square has a value of 1, what is the value of this piece ($\frac{1}{2}$)?' Repeat for $\frac{1}{4}$ and $\frac{1}{8}$. Now do it for $\frac{3}{4}$ and $\frac{3}{8}$. It may help some children to use the whole square as a base and put the other pieces on top of it.

Ask groups of children to find the pieces that are worth $\frac{1}{4}$, $\frac{1}{2}$, and $\frac{3}{8}$. Check that each group has chosen the correct pieces. Also ask them to represent 1 and $\frac{1}{8}$, 2 and $\frac{1}{4}$, etc.

Tell the children the following problem: If someone puts $\frac{1}{2}$ down and someone else puts $\frac{1}{2}$ down, how much is that altogether? Each group must use the pieces to show what it looks like.

Do similar problems with $\frac{1}{8}$ and $\frac{3}{8}$, and $\frac{2}{4}$ and $\frac{1}{4}$. Check that groups are able to use the pieces correctly. (Keep the fraction squares and pieces as children will need them for other activities and may want to use them to solve their own fraction problems.)

Adding fractions

fraction squares

4B, p. 127

Children work through the exercises using their fraction squares to help them find answers. They should write the problems and answers in their own books. When most children have finished, ask some groups to demonstrate how they worked out the problems.

Mixed numbers

fraction squares

Use the fraction squares to work out some more problems.
Write about 20 different fractions on cards, e.g. 2 and $\frac{1}{8}$, $\frac{3}{4}$, $\frac{4}{4}$, $\frac{7}{8}$, 1 and $\frac{5}{8}$, etc. Children have to put the cards in three piles, e.g. those bigger than 1, those smaller than 1, and those equal to 1.

OBJECTIVE 25.2

Working with fractions

Using the fraction squares, ask the children to put the three squares that are in pieces back together again. Encourage them to do this in different ways, e.g. they might put $\frac{1}{2}$ with $\frac{4}{8}$, or $\frac{3}{4}$ with $\frac{2}{8}$. When all groups have done this give them time to show the others how they did it. Tell the following story: 'I have one whole square and I give away a quarter. What do I have left?' Ask the children to discuss how they could use their fraction squares to represent this problem, and show how this would be written ($1 - \frac{1}{4}$). They can work out the answer ($\frac{3}{4}$). Repeat for $2 - \frac{5}{8}$, and for $1\frac{3}{4}$ - _ etc. If children are still not confident do more problems.

new word

fraction squares, rulers and scissors

Fraction differences

Children work through the exercises using their fraction squares, and write the problems and answers in their own books. When most children have finished ask some groups to demonstrate and explain how they worked out the problems.

fraction squares

4B, p. 128

Make your own fraction subtractions

In groups one child uses the fraction squares to show a subtraction. Other children write down the fraction problem that is represented and work out the answer. The children take turns to represent a subtraction and discuss how they did it.

fraction squares

Fractions and number lines

Draw a large number line starting at 0 and ending at 5. Mark every whole number and the quarters between them. Make a dice with the following signs and fractions marked on each side: $+\frac{1}{4}$, $+\frac{1}{2}$, $+\frac{3}{4}$, $-\frac{1}{4}$, $-\frac{1}{2}$, $-\frac{3}{4}$. Children can play in groups of four, each child with a different coloured counter which they place on $2\frac{1}{2}$ to start the game. Take turns to toss the dice and move whatever fraction the dice lands on either forwards or backwards along the number line (if a - sign is shown they move backwards and if a + sign is shown they move forwards). The winner is the first player to reach either 0 or 5.

large number line, fraction dice, counters

OBJECTIVE 25.3

Half of a kilogram

Hold up the one kilogram weight and ask how many grams you need to balance it. Write this on the board as 1 kilogram = 1 000 grams. Ask how many grams are in $\frac{1}{2}$ a kilogram (500 grams). Choose some children to tell how they worked this out. Discuss $\frac{1}{4}$ of a kilogram and $\frac{3}{4}$ of a kilogram. What can the children tell you about $\frac{2}{4}$ of a kilogram? How many kilograms would 1 500 grams be?

a one kilogram weight or a common store item weighing one kilogram

Half of an hour

Ask how many minutes are in an hour. Write this on the board as 1 hour = 60 minutes. Ask how many minutes there are in half an hour, quarter of an hour, and three quarters of an hour. What can the children tell you about two quarters of an hour? How many minutes are there in $1\frac{1}{4}$ hours? What activities in school take half an hour? What games can be played in half an hour?

Half of a minute

Ask if anyone knows how many seconds there are in a minute. Write the answer on the board as 1 minute = 60 seconds. Ask how many seconds there are in half a minute, quarter of a minute, three quarters of a minute, and $\frac{1}{10}$ of a minute. Ask the children to identify activities that take seconds such as blinking an eye or sneezing.

Half past, quarter past, quarter to

Look at Frame A. Ask what time is shown on the first clock (quarter to three). Ask, 'How many minutes before 3 o'clock is shown?' Look at the second clock and ask what time it shows (half past five). Ask, 'How many minutes after 5 o'clock is it?' and 'How many minutes until 6 o'clock?' Continue in the same way with the other clocks.

4B, p. 129

Parts of a kilogram

Children look at each drawing in Frame C and calculate what fraction of a kilogram is shown. They write the fractions in their own books. Check as they are working.

4B, p. 129

Parts of a minute

Do some activities to show how long some of the short times are. For example, how many times can you jump in $\frac{1}{10}$ of a minute? How many times can you clap your hands in one minute?

Fractions of an hour

Using the clock ask children to work out what fraction of an hour has passed if the long hand is on 1 ($\frac{1}{12}$). Do the same for each 5 minute interval around the clock ($\frac{2}{12}$ or $\frac{1}{6}$, $\frac{3}{12}$ or $\frac{1}{4}$, $\frac{4}{12}$ or $\frac{1}{3}$, $\frac{5}{12}$, $\frac{6}{12}$ or $\frac{1}{2}$, etc.).

demonstration clock

Parts of a litre

Children look at each drawing in Frame B. For each drawing, children write the fractions of a litre shown in their books. Check as they are working.

4B, p. 129

OBJECTIVE 25.4

Mixed numbers with fraction squares

Ask groups to use the fraction squares to represent $2\frac{1}{4}$. When all groups have done this look at the ways they have shown it. Is there more than one way of representing $2\frac{1}{4}$? There are a variety of ways children could represent this. For example, they could use two whole squares and two eighths; one whole square, two halves, and one quarter; one whole square, one half, and three quarters; etc. Ask groups to think about some other ways they could represent $2\frac{1}{4}$ without using the fraction squares. Give groups time to discuss this before asking each group to share their answers. (Some possible ways are: using a number line; drawing a diagram to show two wholes and a quarter of another; showing nine quarters, etc.) Accept all suggestions that are appropriate. Do this with a different fraction, e.g. $1\frac{5}{8}$.

fraction squares, rulers and scissors

Fraction Grid

Have children copy the grid into their books and fill in the empty spaces. They draw a diagram in the first column, write a mixed number in the second, draw a number line in the third, and write an improper fraction in the last column. Revise what a mixed number and an improper fraction are. Do the first example together. (In the third column they should mark where $2\frac{1}{3}$ would be on the number line, and in the last column they should write $\frac{7}{3}$.)

4B, p. 130

UNIT 26

DECIMALS

This topic builds on fractions, particularly recognising fractions of shapes, and fractions as a number line.

ENGLISH WORDS/EXPRESSIONS USED

decimal, tenths, thermometer, decimal point

MAIN IDEAS

- **One tenth can be written as 0.1**
- **decimals are used in numbers**
- **reading, writing and recognising decimal fractions**

OBJECTIVES

The children will:

26.1 recognise and write decimal fractions
26.2 revise the use of decimals to represent tenths
26.3 read and interpret decimals where they are used in the world

SPECIAL METHODS AND STRATEGIES

It is important for children to model tenths in many different ways to develop their understanding of decimals.

OBJECTIVE 26.1

Meeting tenths

Choose ten children to come and stand at the front of the room. Ask three of the children to kneel down. Ask what fraction of the group of ten children is kneeling ($\frac{3}{10}$) and what fraction is standing ($\frac{7}{10}$). As they say each fraction write it on the board. Have the ten children stand again and ask five of them to turn their backs to you. Again ask the others to say what fraction is facing you ($\frac{5}{10}$) and what fraction is turning away from you ($\frac{5}{10}$). Write these fractions on the board too. Choose a different group of ten children and do similar activities, making sure you write each fraction on the board.

Look at the fractions that are written on the board. Ask children to tell you something that is the same about all of them. (They are all **tenths**.) Tell the children that there is another way to write these fractions and ask if anyone knows how. Show them how each fraction is written as a **decimal**. For example, $\frac{3}{10}$ can also be written as 0.3, and $\frac{7}{10}$ as 0.7. Choose some children to come out and write some of the tenths as decimals. Explain that 0.3 is said as, 'zero point three' or 'nought point three'. Practise saying some more decimals like this. Tell children that the point is called the **decimal point** and explain its meaning. (To separate the whole numbers from the fractions or parts of a number.)

Treasure Cards

a supply of small cards divided into 10 segments, one dice for each group, treasure items

Hold up a card like this.

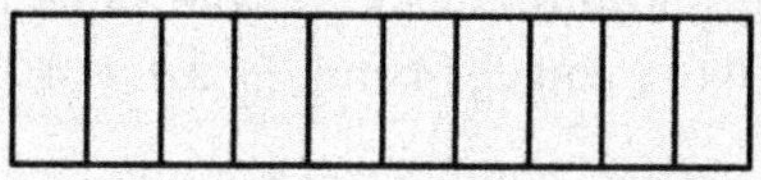

Say that each card, when it is covered with treasure items, is worth 1. Ask how much one section of a card is worth. ($\frac{1}{10}$ or 0.1) Ask about 3 sections, then half the card, asking children to say the part as a common fraction and a decimal fraction each time. Choose three children to demonstrate a game with you. In the centre of the group place some cards and some treasure items. In turn children throw the dice, and pick up that number of treasure items to place on a card, and each time say what decimal fraction they now have on their card. When someone has filled their first card they must start a new one, saying their new decimal tally so as to include any completed cards, e.g. 1.2 or 4.5. Put children in groups of four to play the game. The person from each group who ends up with the largest total number is the winner. Ask different children to use decimal numbers to tell you how many cards they have in front of them at various times.

Decimal shapes

Children should write what the shaded part of each diagram is both as a common fraction and as a decimal fraction. Check as children are working.

4B, p.131

Decimal concentration

Make a *Concentration* game using common fractions and decimal fractions, e.g. $\frac{1}{10}$ and 0.1, $\frac{2}{10}$ and 0.2, $\frac{3}{10}$ and 0.3, etc.

OBJECTIVE 26.2

Decimal fractions of shapes

On the board draw a rectangle divided into ten equal parts with three of the parts shaded. Ask what part of the rectangle is shaded ($\frac{3}{10}$). Ask a child to write this as a decimal (0.3). Repeat for the part of the rectangle not shaded. Draw another rectangle on the board with one of the ten parts shaded and repeat as for the first shape.

Decimals on a line

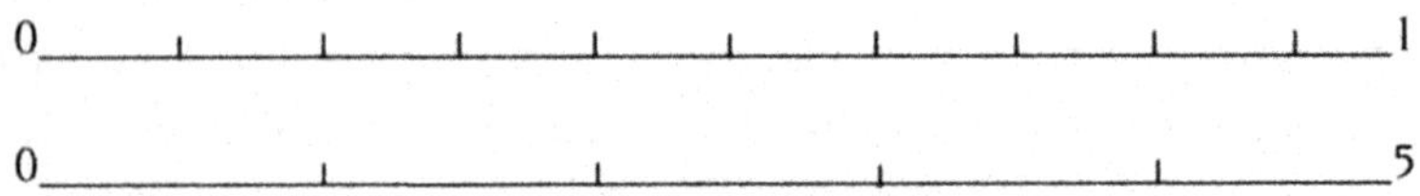

Look at the first number line drawn on the board. Ask what the marks represent (tenths). Choose a child to write $\frac{5}{10}$ above the correct mark on the number line. Choose another child to write this as a decimal below the $\frac{5}{10}$ mark on the number line. In the same way choose other children to write in the other fractions and decimals. Look at the second number line on the board. Point to the mark that represents 3 and ask children what it is. When they tell you write it above the mark. Do the same for the following numbers: 2, 2.5, 0.6, 1 and $\frac{3}{10}$, 4 and $\frac{1}{10}$, 2.9, 1, 4.7, etc.

two number lines drawn on the board

Bottle top decimals

Put children in groups of four and give each group ten bottletops. In turn, each child picks up the ten bottletops and then drops them on the floor in front of them. They count how many landed with the label facing up and write this in their books as a decimal fraction. For example, if two of the bottletops landed with the label facing up they would write 0.2. When each child has had their turn compare the decimals and the child with the highest decimal scores one point. If more than one child has the highest decimal they each score one point. Continue until one child scores five points.

ten bottletops for each group

Decimal grid

Children are to complete each row by filling in the missing fractions, e.g. a diagram in the first column, a number line in the second column, a decimal in the third column, and a proper fraction in the fourth column.

4B, p. 132

Calculator decimals

Ask children to count in tenths on the calculator by putting in 0.1 + = = = etc., and discuss the counting as they go.

calculators

Decimal bingo

Play *Decimal Bingo* by writing about twenty different fractions (tenths) and decimals on the board and asking children to choose six of them to fill in on their own 3 × 2 grid. As you call out each number, if children have it they place a counter on the correct square in the grid. The first child to cover their six squares calls out 'Bingo'. Check by referring to the numbers on the board.

counters, 3 × 2 grid

OBJECTIVE 26.3

Thermometer

new word

Show the **thermometer** and ask what it is used for (to tell the temperature of our body). Explain how to read the thermometer. Then choose someone and take their temperature with the thermometer by placing it under the armpit for about three minutes. The temperature should register between 36.6 and 37.3. Remove the thermometer, read it, wipe it with the cotton wool soaked with alcohol and write the temperature on the board. In the same way take the temperature of as many children as time allows. (If you do not have a thermometer discuss body temperatures with the children using the cut-out thermometer. Normal body temperature is 37.0 but anywhere in the range from 36.6 to 37.3 is normal. A person's body temperature can vary throughout a day by 0.5 to 0.6 degrees. Temperatures above 38.0 indicate illness.) When you have taken the temperatures ask questions such as, 'Who has the highest/lowest temperature?', 'Which ones are above 37.0/below 37.0?', 'Whose temperature is higher than Ben's?', etc.

a clinical thermometer or a large cardboard model, alcohol and cotton wool

Temperature

Children should look at the first group of thermometers and write the temperatures shown on each into their own books. When they have done this they should copy the second group of thermometers into their books and show where the liquid would come to in each thermometer if it was the temperature shown below it. Check as children are working.

4B, p. 133

Thermometer graphs

Continue taking thermometer readings for one child throughout the day. Make a graph of the temperatures recorded. At what time was it the hottest/coolest? Work out the difference between the highest and lowest temperatures.

a clinical thermometer

UNIT 27

SPACE: SHAPES AND MAPS

This topic builds on names and properties of shapes, and also on reading maps as a way of giving and receiving directions.

ENGLISH WORDS/EXPRESSIONS USED

symmetrical, line of symmetry

MAIN IDEAS

- **a shape has a line of symmetry if it can be folded on the line an the folded parts match**
- **a map can represent a place**
- **scales are used to show how big things are**
- **a key and symbols are used to label things on a graph**

OBJECTIVES

The children will:
27.1 identify lines of symmetry in regular shapes
27.2 will investigate symmetry in common objects
27.3 read and interpret a plan with particular emphasis on the scale
27.4 read and interpret a map with particular emphasis on the scale
27.5 read, interpret and construct simple maps

SPECIAL METHODS AND STRATEGIES

This topic uses shapes and maps. You need to collect or make the shapes, and make some maps. If you can find a map of your area, it would be ideal to use that.

OBJECTIVE 27.1

Symmetry in shapes

six very large paper shapes

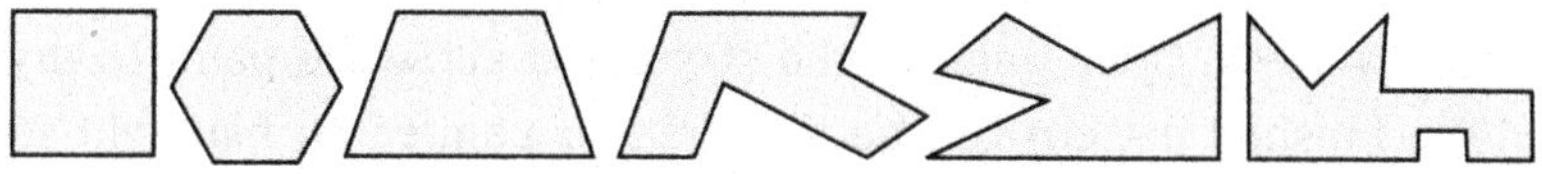

Place six very large paper shapes like these where all the children can see them. Ask what is special about each of the 3 regular shapes. Either you or the children explain that the regular shapes can be cut in half which means that they are **symmetrical**. Invite children in turn to draw a **line of symmetry** on each of the regular shapes. Let the children check what they have done by folding the shapes along the line of symmetry to make sure the two sides match. Ask if the irregular shapes have any lines of symmetry.

Symmetry in Objects

pre-cut paper shapes, blank paper

What other objects in the room can the children see that are symmetrical: tables, windows, chairs, themselves? Invite children in turn to name an object and indicate where its line of symmetry is. Let the other children comment on the suggestions. Show the children one of the pre-cut shapes, squares, rectangles, equilateral triangles, circles etc. Ask how to find the shape's line of symmetry. The children might suggest folding the shape in half. Give out the pre-cut shapes and ask the children to work in pairs to find a line of symmetry for each shape. Have the children draw each shape and mark in the line of symmetry they found for each shape. Share findings and note the different lines of symmetry found for each shape. Challenge the children to find other lines of symmetry for each of their shapes and to draw these on their paper, e.g.

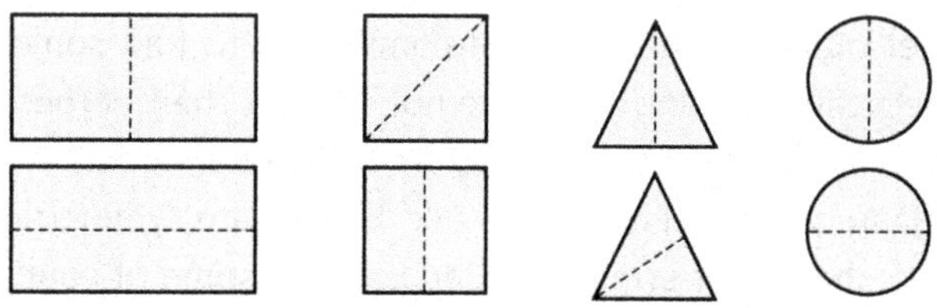

The children share their completed charts and compare how many lines of symmetry they found for each of the shapes. Display their finished charts.

Draw symmetrical shapes

Let the children read through the instructions. Read them aloud in English and vernacular. They can then do the actitvities and share their solutions with the other children.

4B, p.134

Paint Symmetry

Dip a piece of string in paint and lay it in a pattern on one side of a sheet of paper. Fold the paper over to make a line of symmetry and gently press. Open, remove the string and allow the paint to dry. The finished pattern can be coloured or painted to highlight its symmetry.

What is my shape?

Tell the children, I cut out a shape, folded it in half so that the halves matched exactly. Then I folded it in half again so that the halves still matched exactly. Ask the children, 'How many lines of symmetry does this shape have? What might my shape have looked like?' (Answers include all shapes with 2 or more lines of symmetry, e.g. circle, square, rectangle.)

OBJECTIVE 27.2

Symmetry project

a flag book or PNG Primary Atlas, a selection of fruit, a collection of leaves and flowers, or let children collect their own.

The children work with a partner to investigate symmetry. They may choose one of the topics below or suggest their own. At the end they share their findings with the rest of the grade. The topics are:

- Write out your full name in capital letters. What letters in your name are symmetrical? Do these letters have more than one line of symmetry? Can you make up some symmetrical words? (e.g. TOT, BOOK)
- Write the numbers from 0 to 9. Which ones are symmetrical? Can some of these be used to make multi-digit numbers that are symmetrical? (e.g. 8 118)
- Look at the flags of the provinces. Draw some that are symmetrical and some that are not. How many lines of symmetry can you find in your favourite flag? Design your own symmetrical flag.
- Look at buildings. Is your house or your classroom symmetrical? Look at other buildings in your local community to find some which are symmetrical and some which are not. Make a chart of these. Design a symmetrical building (draw a picture or make a model).
- Investigate plants. Many leaves and flowers are symmetrical. Make a chart to show their symmetry. Can you use some of your leaves and flowers to make a symmetrical picture?
- Investigate fruit. Choose as many different fruits as you can. Cut them along their line of symmetry. Are there different ways to do this for each fruit?

Symmetry pictures

Find, or draw, some symmetrical pictures, cut along their line of symmetry and paste each half of the picture on separate pieces of paper. The children choose one and draw the missing part of the picture. Display the finished pictures.

OBJECTIVE 27.3

Map with scale

Ask, 'What is this a plan of?' (a school garden). Give children some activities which focus on the features of the plan, e.g. 'Put your finger on the grade 4 plot?', 'Show where the fruit trees are', etc. Ask children to point to the part of the plan where the scale is marked. Ask what the scale tells us. Ask what one centimetre represents on this scale (1 cm = 10 metres). Estimate, then measure out a distance of 10 metres in the classroom so that children can see how far it is and compare it against one centimetre. Ask what a distance or length of 2 centimetres would represent (20 metres).

rulers (or paper strips) marked in centimetres

4B, p. 135

Length and Width on a Map

Hand out the paper strips marked in centimetres and ask children to measure the length and width of the grade 4 plot (3 cm x 2 cm). Ask them to say what this represents in metres (30 metres x 20 metres). If possible go outside and mark out a plot of this size in the school grounds so that children can see what the real measurements would look like. In pairs children can work out the lengths and widths of each grades' plot, the greens plot, the peanut plot, the cabbage plot, the nursery, etc.

rulers (or paper strips) marked in centimetres

4B, p. 135

Garden size

Ask how much space would be needed in your school ground if you were to make a garden like the one shown on the plan. Ask some pairs to share their answers and tell how they worked it out. Go outside and mark out a garden of the size on the plan. Compare the size of the garden on the plan with your school garden. Is your garden larger or smaller? How big is the grade 4 plot at your school?

metre sticks

4B, p. 135

OBJECTIVE 27.4

Classroom map

Show the map on the board to the children and ask what it is a map of. Ask questions about the features of the room so that all children have a clear picture of it. For example, 'Show where the door is', 'Which wall of our classroom is represented by this line?'. Ask a child to come out to the board and point to the scale. Ask what the scale tells us. (How big things are and how far apart they are.) Ask, 'Can you tell how long our

classroom map drawn to scale (one unit = one metre)

classroom is by looking at the map?' Let children discuss this with a partner and then share their answers and how they worked it out. Have children work out the length or the distance apart of some other classroom features by using the map on the board. Let children check by measuring the real distances in the room

Angoram

The children look at the map of Angoram, a rural centre in the Sepik region. Briefly identify and discuss some of the map features such as the PMV stop, government office, crocodile farm, hotel, etc. Ask questions like, 'If you headed North away from the Sepik River, what large place would you come to?' (Wewak), 'If you walk East along Tobacco Road what will you see?', etc. Point out the scale (1 cm = 50 m). Give each pair of children a ruler (paper strip) marked in centimetres and let them check it against the scale. Explain that on this map one centimetre represents 50 metres, and two centimetres represents 100 metres, etc. Choose a child to measure something on the map, e.g. the distance from the store to the crocodile farm, using the paper strip and then tell the class how many metres this would actually be. Now choose some other children to measure other things on the map or the distance between some of the map features and tell the class the actual distance.

paper strip or ruler

4B, p. 136

Map of the School

Using a centimetre grid children in groups can draw a map of the school. They will need to work out a suitable scale, e.g. 1 cm = 10 m depending on the length of the school buildings and grounds.

centimetre grid

OBJECTIVE 27.5

Map of the National Capital District

Tell the children that this is a map of part of the National Capital District. Discuss the features shown on the map. Now discuss the scale used. Note that the scale used is one centimetre = one kilometre. Ask children a series of questions about the distance between the key features marked on the map.

4B, p. 137

Map of my house

The children draw a map of the room where they sleep. They decide what symbols to represent the features of the room. They may use a scale for their map. The children can show their work to the whole class.

UNIT 28

LONGER LENGTHS

This topic builds on estimating and measuring lengths in metres and centimetres.

ENGLISH WORDS/EXPRESSIONS USED

armspan, pace

MAIN IDEAS

- **length in metres can be used to compare distances**
- **body parts can be used as standard measures**

OBJECTIVES

The children will:

28.1 estimate, then measure the dimensions of various sports fields in metres

28.2 estimate, then measure and compare various body measurements in metres and centimetres

28.3 estimate, measure and compare the distance around trees

28.4 determine the length of their pace as a fraction of a metre and use it for estimating length

28.5 estimate and measure various South Pacific Games records

SAMPLE ACTIVITIES

OBJECTIVE 28.1

Volleyball court

If your school has a volleyball court (or similar) then children estimate the length and width of the court. They write down their estimates and then measure to see how close they were.

a volleyball court or space to mark out a court, metre sticks

Sports fields

Look at the plans of the sports fields and discuss the various measurements asking questions such as, 'Which sports field has a side that is 120 metres long?', 'Which sports field is the widest?', 'About how many paces would it take to go from one end of the volleyball court to the other end?', 'Is our volleyball court the same size as this one?', etc.

4B, p. 138

Measuring out a sports field

Put children in groups of three or four. They can do one of the following. Choose one of the sports fields in the book and use the measurements given there to mark it out in their school grounds. Choose one of the sports fields at their school and draw a scale plan of it writing in each measurement as they find it. Check their work.

4B, p. 138

More sports fields

Find measurements of other sports fields and compare them with the ones you know. Have the children work out the perimeter of the volleyball court.

OBJECTIVE 28.2

Armspans and heights

new word

Choose one child to hold their arms out wide. Ask the other children to estimate which is longer, their height or their **armspan**. If possible mark their height and armspan against the board. Now choose a child to come out and use a metre stick to measure the height of the child as marked on the board and to write this as metres and centimetres. Do the same for their armspan. Compare the two measurements. Which is longer? Put the children in groups of four and ask them to repeat the activity to see if what they found out applies to everyone. They should write down all measurements in metres and centimetres. Give groups time to share their findings.

paper strips

Body measurements

Children work in pairs to find the longest body part in each group of three pictures. They write each measurement in metres and centimetres. Check that they are measuring correctly.

4B, p. 139

Half your height

Ask each child to use their height measurement to work out how tall someone would be who is half their height. Ask some children to tell how they did it. Now ask children how old they think someone who is half their height would be. Write down their estimates and discuss how they could find out. Tell each child to find someone in their village who is half their height and ask them how old they are. Compare findings.

Double your height

Work out how tall you would be if you were double your height. Make a life-size model of a child in your class by using scrap paper or fabric to stuff some clothing with. Make the model the same measurements as the child.

OBJECTIVE 28.3

Distance around trees

ten trees, labels from 1 to 10, string or vine, metre sticks

Take the children outside and point out the ten trees that they are going to be working with. Stick a numbered label on each tree. Choose one tree and by putting your arms around it show them which part to measure. Ask them to work with a partner to estimate which of the ten trees has the largest distance around. They then estimate which has the next largest distance around and so on down to the tree with the smallest distance around. Discuss. When children have done this ask them to suggest how they could measure the distance around each tree. They should realise that they will need something flexible to go around the trees. Give children string or vine and allow them to use it to find the distance around each tree and then use one or more metre sticks to find the measurement of the string. Children should write the measurement of each tree next to its number in their books. When all children have finished return to the classroom and check their measurements against their estimates. Was the tree with the biggest distance around it the one they estimated to be the biggest? etc. Discuss with the children how to work out the difference between the biggest tree and the smallest tree. Can the children find something in the classroom that is as long as the distance around the biggest tree? Can they find something as long as the distance around the smallest tree?

OBJECTIVE 28.4

Measuring your pace

new word

Make a line on the ground outside to use as the starting point. Ask each child to estimate, then **pace** out ten paces starting from the line you have marked and to make a smaller mark where they finish. They can now use their metre stick to measure their distance from start to finish to the nearest ten centimetres. When all children know how far ten of their paces takes them return to the classroom. Children should write down this distance to the nearest ten centimetres, e.g. 9 metres and 40 centimetres, or 940 centimetres.

When they have done this put them in groups and ask them to discuss how they could use their measurement for ten paces to work out the length of one pace. (They will need to divide the measurement for ten paces by 10.) Accept any ideas that work. As a class work out some ways of doing this, e.g. use a calculator; change the distance to centimetres and then divide by 10; etc. Write some of the distances for ten paces on the board and do some with the children until they feel confident. Now ask each child to work out their own but help those children who need it. Use the pace length to find some other distances around the school building. Keep calculations within a manageable range.

counters, metre sticks

Measuring length in paces

Ask children to estimate, then find how many paces it takes to walk the length of the classroom, the width of the classroom, etc. When all children have done this discuss how you can use the measurement of one of their paces to work out the lengths of these things. If a calculator is available use it to help work out the lengths. If you do not have a calculator show children how they can keep adding the length of one pace, or multiply the centimetres by the number of paces. Help children to do this. Compare the final measurements with those of other children. How accurate are these measurements? Why?

a one metre stick marked in ten centimetre segments

Big paces

Children should look at the pace length of each child. Ask, 'Who would take the most steps to reach the tree?', 'If Kilala takes 25 steps to reach the tree how far away is it?', etc. Discuss how to work out these calculations and give children time to do it.

4B, p. 140

OBJECTIVE 28.5

South Pacific Games records

Discuss what event is shown in each picture and talk about what each competitor has to do for that event. Mark a starting point on the floor or on the ground outside. Ask children to stand where they think the women's South Pacific Games record would be for a woman competing in the long jump (7.49 m). In pairs have children measure their estimates and write the distance down in their books in metres and centimetres, and centimetres only. Write the record on the board in metres and centimetres and ask the children to tell you what this is using centimetres only. Now measure it out on the floor or the ground. How close was their estimate? Repeat for the men's triple jump (15.66 m), the women's javelin (49.68 m), and the men's discus (51.32 m). For the men's high jump (2.05 m) you will need to mark a distance upwards. It may be better to do this as a class using a long stick to indicate the height. When you have written the South Pacific Games record on the board for each of the events, put children in pairs and ask them to use their metre stick to find something at the school which is the same length or height as each South Pacific Games record. When children have had time to do this choose some pairs to tell about what they found.

a one metre stick marked in ten centimetre segments

4B, p. 141

Class records

Let children do a long jump or a hop, step and jump, and compare their efforts to the South Pacific Games records. How much further do they need to jump to equal the South Pacific Games record? Let them compare the South Pacific Games records for these events with some local records.

a one metre stick marked in ten centimetre segments

UNIT 29

CAPACITY

This topic builds on *Teacher's Resource Book* 4, Unit 15, where children were introduced to the $\frac{1}{2}$ litre as a unit of capacity/volume and compared capacities using several different methods.

ENGLISH WORDS/EXPRESSIONS USED

millilitres, drops

MAIN IDEAS

- **that millilitres are used for measuring small volumes**
- **that litres and millilitres can be used to compare the capacity of containers**

OBJECTIVES

The children will:

29.1 estimate $\frac{1}{4}$ of a litre, and be introduced to mL

29.2 use mL as a way of measuring capacity

29.3 compare the capacity of various containers and use this as a basis for exploring mL

29.4 convert measurements written in litres to millilitres and vice versa

29.5 use estimation and counting as a way of developing a sense of capacity

SPECIAL METHODS AND STRATEGIES

As with other measurement units, it helps if you use practical examples, and activities from the experience of the pupil or which are available in the local area.

OBJECTIVE 29.1

Parts of a litre

Hold up the one litre container and ask children to estimate how much it holds. After some children have estimated ask them how they could check. Now ask them to estimate where one quarter of a litre would be on your container. That is, if it was filled to a quarter of a litre, where would it be filled to? Choose some children to use a pen to mark on the container where they think this would be. Write their name or initials next to the mark and put the container to one side for the time being. Put children in groups and give each group a litre container and a marking pen. Tell the children that they are to each estimate where a quarter of a litre would be on their container and place a mark to indicate their estimate. They then discuss how they could find out whose estimate is closest. When groups have had time to discuss this let each group share their strategies. (Two possible ways are: find a container that holds exactly a quarter litre, fill it and pour it into the litre container; share the litre equally between four containers by using, for example, cupfuls, and then pour one of these containers back into the litre container.) Now let each group try their strategy. When they have done so let them report back to the others. Which strategies worked? Whose estimate was closest?

Let children choose one of the strategies that worked to check the container you used at the beginning of the activity. Who estimated correctly?

different one litre containers, various containers including $\frac{1}{4}$ L, marking pens

Estimating litres

Children work as a class to do this activity. They look at each container to see how much it holds. Then they look to see how much of the container is filled and decide what amount is. For example, the first container holds one litre so it looks to be filled to a $\frac{1}{2}$ litre. As you do each of these ask questions such as, 'What is another way of saying $\frac{1}{2}$ litre?' (500 mL) 'What does 1 000 mL mean on the third container?' (1 litre) 'Which containers hold exactly $\frac{1}{2}$ a litre?' etc.

4B, p. 142

Finding $\frac{1}{4}$ of a litre

Find some other containers that hold exactly $\frac{1}{4}$ of a litre. Ask questions like, 'How many $\frac{1}{4}$ L containers hold the same as a 1 L container?', 'How could you mark $\frac{1}{4}$ L on a $\frac{1}{2}$ L container?'

OBJECTIVE 29.2

Using millilitres

new word

Hold up a one litre container and ask the children to tell you how many **millilitres** are in one litre. (1 000 mL) Write this on the board using mL to show how many millilitres. Say, 'If one litre is 1 000 mL, then how many millilitres will be in half a litre?' (500 mL) Write this on the board too. Now ask, 'How many millilitres will be in one quarter of a litre?' (250 mL) Also write it on the board. In the same way establish how many millilitres are in three-quarters of a litre. (750 mL)

a one litre container

More or less than 500 mL

It is a good idea to do this activity outside as children will need to use water to fill their containers. Put children in groups and give them a one litre container, a 500 mL container, and a variety of containers that hold less than one litre. Be careful not to tell children what any of the containers other than the litre one holds. Tell them that their task is to find a container that holds exactly 500 mL, and then to put their containers in two groups - those that hold less than 500 mL and those that hold between 500 mL and one litre. Give groups time to do this. Once children have sorted their containers into two groups they can try to work out the capacity of each one. For example, do any of their containers hold 250 mL? Are there any that hold 750 mL? What might the other containers be? When groups have done this give them time to report to the other groups about what they found out.

Make a collection of containers that hold exactly 250 mL or 750 mL. Visit a trade store to see what is sold in amounts less than 250 mL.

one litre and 500 mL containers, various containers smaller than one litre

Estimating capacity

Give each child some counters. Ask them to place counters on certain pictures. For example, 'Put counters on all the containers that hold more than 500 mL.', 'Put counters on all containers that hold $\frac{1}{4}$ of a litre.', and 'Show the containers that hold less than 250 mL.' Check that children are placing counters on the correct pictures.

counters

4B, p. 143

OBJECTIVE 29.3

Drops

new word

Put a teaspoon, a cup, and a litre container where all children can see them. Ask the question, 'How many **drops** of water are needed to fill the teaspoon?' Invite some children to estimate and then use the eye dropper (or a slow dripping tap) to do it. Write the answer on the board. How close were their estimates? Now ask, 'How many

teaspoon, cup, litre container, eye dropper

teaspoons of water are needed to fill the cup?' Ask the children to estimate and write their estimate down in their books. Put the children in groups and give them a teaspoon and a cup. They are to work out how many teaspoons the cup holds and write this down next to their estimate. Whose estimate was close?

Cups in a litre

Give each group a litre container. Ask children to estimate how many cups are needed to fill the litre container. They should first estimate and write it in their books. Then they should do it and write the actual number of cups in their books next to their estimate. How close was their estimate?

teaspoons, cups, one litre containers

Drops in a litre

When groups have finished, as a class use the information to work out how many drops of water would be in one litre. Then ask children to suggest how they could work out how many millilitres are held by the teaspoon. Discuss the suggestions, then ask children to work in the same groups as before and use one of the suggested strategies to find how many millilitres a teaspoon holds. Give groups time to report on their findings.

teaspoon, cup, litre container, eye dropper

Filling a container

The children's task is to find a combination of containers that when put together would fill the 750 mL container at the top of the page. They should write their choices in their own books. Is there more than one possible combination? They are to do the same thing for the one litre container at the top of the page. How many different combinations can they find for it?

4B, p. 144

Small containers

Make a collection of containers that hold small amounts. What is the smallest they can find? Visit a trade store to see what is sold in small amounts. Is there a reason for some things being in small containers?

OBJECTIVE 29.4

mL in a litre

Hold up the two litre container and ask children to tell you how many millilitres it holds. If necessary remind them that one litre is 1 000 mL. On the board write the children's answers as:

1 litre = 1 000 mL
2 litres = 2 000 mL

Continue this chart until you reach 10 litres. Ask children to copy it into their own books.

a two litre container with the capacity marked

Difference in capacity

Each container has its capacity written on it in litres, or litres and fractions of litres. Children are to write the capacity of each container in millilitres. Check as they are working. Use the containers to ask other questions such as, 'What is the difference in millilitres between the______container and the______container?' Use a variety of words for difference such as, 'Which is more and by how much?', 'How much more does _____hold than_____?'

4B, p. 145

Ordering capacity

Put children into groups and give each group some of the containers. Groups should first put their containers in order of capacity. Then if the capacity of the container is written in millilitres children must write it as litres, e.g. 1 250 mL = 1 litre and 250 mL, or $1\frac{1}{4}$ litres. If the capacity is written in litres children must write it in millilitres, e.g. 2 litres = 2 000 mL. Put all the class containers together and ask children to find a group of containers that together hold $1\frac{1}{2}$ litres. Ask another child to do this in a different way. Continue until the children have found as many ways as possible. Repeat for $2\frac{1}{4}$ litres, $\frac{3}{4}$ of a litre, etc.

various labelled containers larger than one litre, various labelled containers smaller than one litre

Solving capacity stories

There are a number of problems on the page. Ask the children to solve the problems and to write their answers in their books.

4B, p. 146

OBJECTIVE 29.5

Grains of rice

Put the children in groups and give each group some rice. In turn each child in the group should take a handful of rice and estimate how many grains of rice are in their hand. They should write their estimates in their books. Have a class discussion to find out what some of the estimates were. What is the highest estimate? The lowest estimate? Are there any estimates that are unlikely? Ask how they could find out how many grains of rice are in a handful. Talk about ways of counting to make the task easier, e.g. finding how many grains are held in a spoon or similar and then counting how many spoonsful are in your handful. Children should return to their group and do this. When all groups have done it compare their answers. Are the answers the same? Why are they different? How close are they? How close was their estimate?

rice or small seeds

Handfuls in a litre

Ask groups to estimate how many handfuls of rice it would take to fill a one litre container. Then give each group a one litre container to do it. How close was their estimate? When they have found this they should work out how many grains of rice are in a litre. Share the answers with other groups, discussing reasons for any differences.

one litre containers, spoons, small scoops, rice or small seeds

Shopping lists

Use the shopping lists on the page to work out which family is buying the greatest total capacity and which family is buying the smallest total capacity. When children have done this they should find the difference between the largest and smallest. Then ask, 'If we put all the milk in one container what size would we need?' Give children time to work this out. Ask more similar questions.

4B, p. 147

Differences in capacity

Use containers in the classroom to do activities where children have to find the difference between capacities. For example, 'What is the difference in capacity between this 2 litre cordial bottle and this 750 mL container?' or 'How much more does this large juice container hold than this small one?'

UNIT 30

PERIMETER AND AREA

This topic builds on estimating and measuring length, measuring length of lines on a grid, understanding that area is covering.

ENGLISH WORDS/EXPRESSIONS USED

irregular

MAIN IDEAS

- **perimeter is the distance around and we can measure it in centimetres or metres**
- **area is covering, and we can measure it in square centimetres or square metres**

OBJECTIVES

The children will:

30.1 draw and calculate perimeters of shapes which are drawn on grid paper

30.2 use square centimetres for calculating the area of the shapes of letters

30.3 calculate the area of shapes requiring the use of half squares

30.4 draw shapes with a given area

30.5 compare area and perimeter

SPECIAL METHODS AND STRATEGIES

Show children how to record the perimeter of an object, e.g.
Perimeter = ____ centimetres or P = ____ cm
They should also learn to record area in the following way:
Area of rectangle = ____square centimetres or A = ____square centimetres

OBJECTIVE 30.1

Perimeter of a rectangle

On a grid of squares (about 10 x 10) on the board draw a vertical line from one intersection (a point where the lines meet) to another. Ask how many units long the line is. (Count the squares, not the intersections). Repeat this for other lines. Draw a rectangular shape on the grid lines. Ask children to estimate the perimeter of this shape. As a class count the number of units around the outside of the shape. Repeat for other rectangular shapes.

a grid of squares drawn on the board

Rectangles of a given perimeter

Tell the children that a rectangle has a perimeter of 24 units. Ask them to think about what the rectangle might look like and draw it on the grid paper or in their books. When children have finished compare their rectangles. How many different ones are there? Can anyone think of any more? Have children draw some more rectangles with perimeters of 18 units, 20 units, and 12 units.

grid paper, pencil, ruler

Large and small perimeter

Tell the children to work with a partner. They should, together, estimate which shape has the largest perimeter and which shape has the smallest perimeter and write this in their books. Now children should calculate the perimeter of each of the rectangles and write it in their own books. Did they estimate correctly? Do any of the shapes have the same perimeter as each other? Ask pairs of children to discuss with each other how they worked out the perimeter of each rectangular shape. Choose some children to report their group's method to the class. (Did anyone add the length and the width together and then double it? This is the easiest way to do it but it is better if children discover this by themselves, rather than the teacher telling them to do it this way.)

4B, p.148

Perimeter of other shapes

new word

Draw a rectangular shape on the grid on the board. Ask children to work with a partner to calculate the perimeter of the shape. Ask some to tell you their answer, and explain or demonstrate how they did it. Praise successful methods and encourage any children who are having difficulty to try out these methods. On the grid draw another shape which is **irregular** but which follows the grid lines,

a grid

e.g.

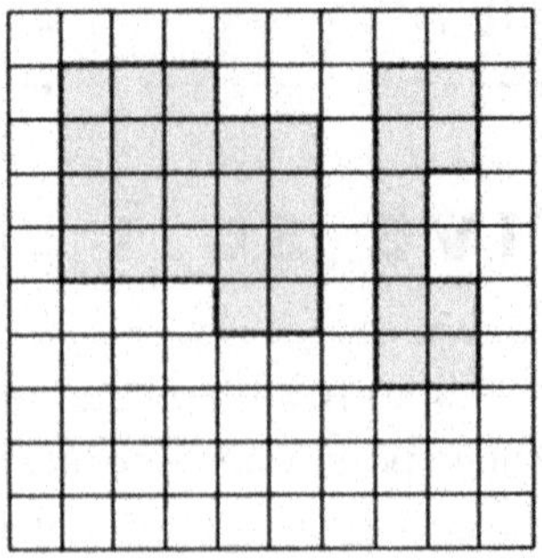

Ask pairs of children to talk about how they would find the perimeter of this shape. Choose some pairs to share their answer and method. Draw another irregular shape which follows the grid lines. Repeat, checking that all children understand and can find the perimeter of a shape.

Estimating perimeters

Children estimate which shape has the largest perimeter and which shape has the smallest perimeter. Then they calculate the perimeter of each shape and write it in their own book. Did they estimate the largest/smallest perimeter correctly?

4B, p. 149

Finding a given perimeter

Working with a partner children use string to try to find something in the classroom which has a perimeter of one metre or 100 centimetres. Share the findings with the rest of the children. Did anyone find something which has a perimeter that is exactly 100 centimetres? Can they find more than one thing?

string

Perimeter of objects using units

Using measuring instruments marked in metres or centimetres, children find the perimeter of some regular shaped objects, e.g. books, maps, tabletops, the classroom.

measuring instruments

OBJECTIVE 30.2

Area on a grid

On the grid on the board draw the letter 'U' using the squares of the grid, e.g.

a grid

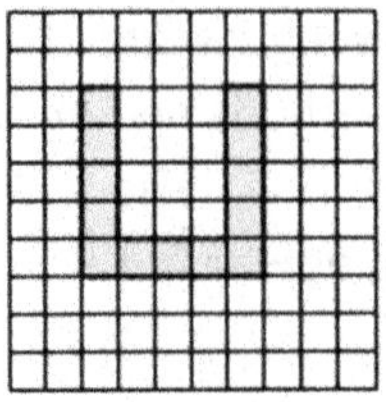

Tell the children that each square on the grid represents a square centimetre. Ask how many square centimetres are needed to cover the letter 'U'. Put the children in groups to discuss how they might calculate the answer. Share the strategies and decide which are most efficient.

Area of letters

They are to estimate which of the letters has the largest area and which has the smallest area. Then they calculate the area of each letter and write it in their own book. Did they estimate correctly? When most children have finished discuss how they worked out the area of each letter.

4B, p. 150

Area of words

Put the following words on the board: FEET, HAT, TALL, and LIFT. Ask the children to work out how much area each of these words would take up. Each word is made up of letters shown on page 150 so children should use the areas they found for each letter to calculate the area of each word. Check while children are working. Children should work in pairs to make some more words from the letters on page 150. Each letter can be used more than once in a word. Which word has the largest area? Ask children to find a word with a greater area than any of the words found so far using only the letters on page 150 and the letter 'U'.

4B, p. 150

OBJECTIVE 30.3

Area using part units

On the grid on the board draw the letter 'C' with straight sides, but with the corner squares cut in half as shown.

a grid

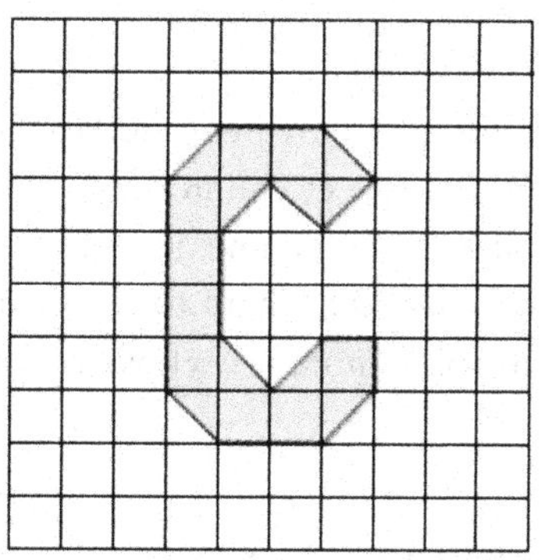

Tell the children that each square on the grid represents a square centimetre. Ask the children how many square centimetres would be needed to cover the letter 'C'. Put the children in groups to discuss how they might calculate the answer. Share the strategies and discuss the difficulties, particularly how they might work out the half squares. On the same grid draw the letter 'P' as shown.

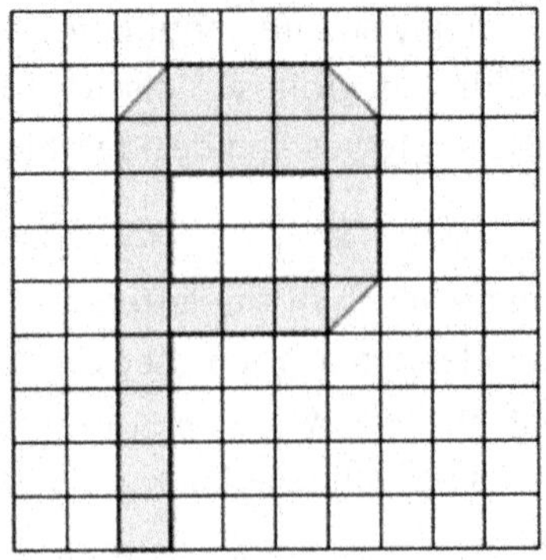

Have the children calculate its area using one of the successful strategies discussed above. Share the answers.

Area of more letters

Estimate which of the letters has the largest area and which has the smallest area. Now they should calculate the area of each letter and write it in their own book. Did they estimate correctly? When most children have finished discuss how they worked out the area of each letter.

4B, p.151

More area of words

Put the following words on the board: BOSS, BOX, POOR, and ROD. Ask the children to work out how much area each of these words would take up. Each word is made up of letters shown on page 151 so children can use the areas they found for each letter to calculate the area of each word. Check while children are working.

4B, p.151

Area of your own words

Make a list of words and their areas. Add to the list as children make words with larger areas. Let children use the letters on pages 150-151 to find the area of their name. If a letter in their name does not appear on either page help them to work out its area so they can calculate the area of their name. Whose name has the largest area?

4B, p.150 and p.151

OBJECTIVE 30.4

Drawing shapes of area 12 sq. cm

Children trace the grid into their own books or onto some paper. When they have done this they should draw a shape on the lines of the grid (no half units) which has an area of 12 square centimetres. Swap shapes with a partner and check that the area is 12 square centimetres. Choose a child to come out to the board and use the

paper for tracing

4B, p.93

grid there to draw their shape. Ask another child who has a different shape to do the same. Repeat for another child. Ask the children to use their own grid to draw another shape (again, they are to use whole units) which has an area of 12 square centimetres. When children have finished give them time to look at other children's shapes so they can see the range of possible shapes.

Perimeter challenge

Now ask the children to measure the perimeter of their shape and list some of their responses on the board. Discuss the result, e.g. how can all the shapes with the same area have different perimeters? Now challenge the children to do the following. For a shape with a fixed area of 12 square centimetres draw the shape with (a) the largest perimeter, and (b) the smallest perimeter.

Drawing other shapes of area 12 sq. cm

Ask the children to draw a shape which has an area of 12 square centimetres using some half units. Allow time to share the completed shapes, and to check that they are 12 square centimetres. When children have done this they may use their grid to draw some more shapes with the same area.

Calculating area

The children estimate the area of each shape and write it in their own books. They should then find each area by counting the square centimetres, and write this in their books too. How close was each estimate to the actual area?

4B, p. 152

Finding objects of a given area

Find some objects in the classroom that have a base area of, say 60 square centimetres. Use these objects to estimate the area of other objects. Make a list of objects which have an area less than 10 square centimetres.

OBJECTIVE 30.5

Comparing area and perimeter

Ask the children to turn back to where they recorded the perimeter of these shapes. Ask them to rewrite the perimeters in order from smallest to largest. Now ask them to calculate the area of each of these shapes and write these in order from smallest to largest. Ask the children to compare the two lists and discuss what they notice with a partner. Choose some children to report their findings back to the class.

4B, p. 149

UNIT 31

SETS

This topic builds on work in *Teacher's Resource* Books 1, 2 and 3, in which children were asked to sort, classify and describe collections of objects and properties of individual objects. This topic makes this more formal, introducing the words set and elements.

ENGLISH WORDS/EXPRESSIONS USED

elements, members, overlapping, set

MAIN IDEAS

- **a set is a collection of objects**
- **elements of a set have something in common**
- **sometimes elements can belong to two sets**

OBJECTIVES

The children will:

31.1 discuss similarities and differences, and will use the language of sets to describe groups

31.2 list the elements of sets

31.3 identify objects that belong in two different sets

31.4 identify objects which belong in two different sets and use this information to help solve problems

SPECIAL METHODS AND STRATEGIES

The key in this topic is to encourage the children to talk in English. They will have their own ideas, which will be valid but different. Support all contributions.

OBJECTIVE 31.1

Making sets

a collection of classroom objects

Have the children sitting in a circle. Give each child one of the objects from the collection. Say that their task is to put these objects into groups or **sets** that belong together. Ask a child to start by placing their object in the centre. Then have each child in turn place their object either in a set already in the centre or separate from them to indicate a new set being formed. When all the objects have been placed in sets ask, 'What is the same about all the **elements** of this set?' Explain that a set is any group of things which are connected in some way. The items in a set can be called 'elements' of the set or **'members'** of the set. Discuss each of the sets in this way.

new word

Sorting letters

Tell the children to write the capital letters of the alphabet across the top of a page in their books. They must then sort the letters into two sets, crossing out each one as they include it in a set. For example, in one set they might put all the letters made with one pen stroke, e.g. J, I, L and in the other they put all the letters made with more than one pen stroke. When the children have finished, ask someone to read out the elements of their first set. Let the others decide what is the same about these letters. Then let them work out what letters must be in this child's other set and why.

Sorting food

4B, p. 153

Tell the children to look at the food in Frame A. Tell them that they are to copy the food into their books, grouping it into two sets. When they have finished ask some children to read out the sets they made and say what is the same about the elements in each set. Discuss the sets in Frame B and then Frame C. What is the same about the elements in each set?

What's My Rule?

Put 4 to 5 children into 2 sets without telling what your rule is, e.g. hair length, footwear, shirt colour, etc. Invite children one by one to join one of the sets, but they must not speak or say what they think the categories are. If a child joins the wrong set, let them wait until all the other children have had a turn before trying again. When the grouping is complete choose a child from each group to say what is the same about all the elements of their set.

Sorting games

Ask each child in turn what their favourite game is and list these on the board. Put the children into groups of four and let them discuss what sets these games could be put into, e.g. board games, team sports, play-acting games, outdoor games, etc. When they have had time to do this, invite each group to report their discussion and their list to the rest of the class, saying why the members of each set belong together.

OBJECTIVE 31.2

Sets of people

Ask a child to name the set of people that live in their house. Choose other children to answer similar questions such as, 'Tell me the members of the set of teachers at this school.', 'Tell me the members of the set of children who are wearing blue today.' or 'Tell me the members of the set of children that come from _____ village.'

List the elements

4B, p. 154

Tell the children to look at the numbers in Frame A. They write down the members of the set of odd numbers from this group. Check their answers when they have finished. The children list the elements of the set of triangles, the elements of the set of animals with six legs and the elements of the set of animals that live in trees.

Guess my set

Tell the children to think of a set of things and write the names of 10 elements of their set in their books, without telling anyone what these elements have in common. When they have done this, choose a child to read out the first three elements in their set and let others guess what these items have in common. If they cannot guess easily, ask the child to read out two more items from their set and continue until the class has worked out what the set is based on. Choose several more children to read out their list.

Write as many elements as you can

Put the children into groups of four and have competitions about how many elements they can think of for a particular set, e.g. red things in this room. Give them time to write down as many members as they can think of. Let them compare their answers and see who thought of the most elements for the set. Provide a list of topics so they can continue their game.

OBJECTIVE 31.3

In two sets

Have the children sitting in a circle and give out the items in your collection, e.g. some red things, some wooden things and some things that are red *and* wooden, or, some things that you write with, some blue things and some things that are blue *and* are used for writing, for children to hold. Ask the children to help you put the items into two sets. Invite children, in turn, to place their item in either the red set or the wooden set. Let them continue until all the items have been placed. If no child has raised the problem of what to do with an item that could go in either of the sets, e.g. a red pencil, then choose one of these items and say, 'But what about this? It could go in the other set as well.' Discuss what to do with the item and also find the other items that could be placed in either the red set or the wooden set. The children will probably suggest placing these items somewhere between the two other sets. Draw two large intersecting circles on the floor and label one 'Red' and the other 'Wooden'. Ask the children to put the red things in one circle and the wooden things in the other. Discuss where the items that are both red and wooden should be placed. When the grouping is complete, ask a child to make a label for the intersecting set, e.g. 'Red and Wooden'. Point out the advantages of grouping the items in **overlapping** circles instead of making three separate sets. It shows how the elements in the three sets are related to each other.

new word

a collection of items.

More than two sets

Using your collection of items, hold them up one at a time and ask, 'Which set is this an element of?'. Let children place each item where it belongs. Discuss where to put the items that do not belong to either of the two sets used in 'In two sets'. A new set is needed.

some items from *In Two Sets*, some other different items

Which are in both?

Tell the children to look at the numbers in Frame A. Tell the children to draw two intersecting circles in their books so that one circle contains numbers belonging to the two times table and the other contains numbers belonging to the five times table.Copy the numbers to show which ones belong in each set and which ones belong in the intersection of the two sets. Repeat for the other Frames on the page, e.g. shapes with a curved edge and shapes with a right angle, market food and supermarket food, animals that walk and animals that swim. When most of the children have finished, choose some children to say in which part of their intersecting circles they put the numbers in Frame A. Let other children comment. Do the same for the other three Frames on the page.

4B, p. 155

Animal sets

Draw two large intersecting circles on the board. In the first circle write or draw some furred animals such as a dog, a wallaby and a rat. In the intersection of the two circles draw two furred animals that live in trees such as cuscus and tree kangaroos. Ask children if they can explain the diagram and suggest other animals that could be included in the sets already there or placed in the empty set of animals that live in trees.

Make your own sets

Draw three intersecting circles on the board. Choose children in turn to suggest an element and say the set they think it belongs to. Draw or write the element in its set if the suggestion is correct, or put it in the area outside the three sets if it is not. When the children have suggested a number of elements for each set, invite children to make up a name for each set.

OBJECTIVE 31.4

Playing sports

two very large string circles

Put the two large string circles on the ground. They do not have to be overlapping at this stage. Tell the children that they are going to help solve a problem using sets. Say, 'The problem is about a class of 32 children.' (Use the number of children in your class.) 'They have entered a sports competition against some other schools. They need 20 children to play basketball and 18 children to play volleyball. How can we arrange this so that everyone has the chance to play one of the sports?' Ask for 20 children who would like to play basketball to stand in one of the circles and count that they have the right number. Now say, 'Now we need 18 children to play volleyball. We only have 12 children left. What shall we do?' If none of the children suggests it, move one of the circles so they overlap and invite 6 of the basketballers to stand in the area that is also for the volleyball players. Ask children to make up a name for each group, e.g. 'basketballers', 'volleyballers' and 'children who will play both sports'.

Taking trips

State the problem 'We have a boat that will take 24 children out on the lake in the morning and a bus that can take 14 children up to the top of the mountain in the afternoon.' Label one of your circles 'boat' and the other one 'bus'. (Or make up a similar situation to suit your class.) Say, 'Put your hand up if you would like to go on the boat.' Choose 24 children to stand in the 'boat' circle. Now say, 'We have room on the bus for 14 people but there are only 8 children left.' Have these children stand in the 'bus' circle. Say, 'There will be six spare seats on the bus. What can we do?' If no-one suggests it, ask who would like to go on both trips and choose six of these children to stand in the area where the circles overlap to show they are part of both the 'boat' and the 'bus' circles.

Lolly problem

Put the forty cubes in the middle of the circle. Tell the children a story of how the trade store owner had packets of new lollies and he wanted the teacher to give them to the children in his class to see if they liked them. Say, 'Put your hand up if you would like a packet of these raspberry lollies.' Give out the 20 red cubes. Now give out green cubes to the other children in the class and show how many you have left. Ask who would like to try the lime lollies as well. Give out the other 8 green cubes. Those children who tried both the raspberry and lime lollies should join their cubes together. Now tell the children with red counters to place them in one of the circles and children with green counters to place them in the other circle and children with a red and a green counter to link them together and place them in the intersection of the circles. Ask children to say how the circles show how the lollies were shared out.

20 red linking cubes and 20 green linking cubes

Set problems

Choose a child to read the first problem. Choose another child to say in their own words what the problem is. Repeat this for the other problems on the page. The children draw two pairs of intersecting circles in their books and use the circles to help them solve the problems. They discuss their work with a partner as they go. When all the children have finished choose several pairs to draw their solution on the board and demonstrate how they worked it out.

4B, p. 156

UNIT 32

GRAPHS

This topic builds on *Teacher's Resource Book* 3, Unit 3, in which students read, interpreted graphs, recorded information, and conducted surveys. This topic extends this work, and uses one symbol to represent more than one digit.

ENGLISH WORDS/EXPRESSIONS USED

graph, survey, data

MAIN IDEAS

- **information can be read from graphs**
- **one symbol can sometimes be used to represent more than one object**
- **information can be presented on graphs**

OBJECTIVES

The children will:

32.1 use graphs which represent changes over time

32.2 interpret graphs drawn over time

32.3 read and interpret graphs which use one symbol to represent more than one object

32.4 draw graphs using one symbol to represent many objects

32.5 decide on a topic, collect data and record their data in graph form

SPECIAL METHODS AND STRATEGIES

It helps if graphs are relevant to the children. Adapt the information here to your own class. It is important that the children can discuss what they think the graphs mean.

OBJECTIVE 32.1

House lights

Draw a graph on the board like the following:

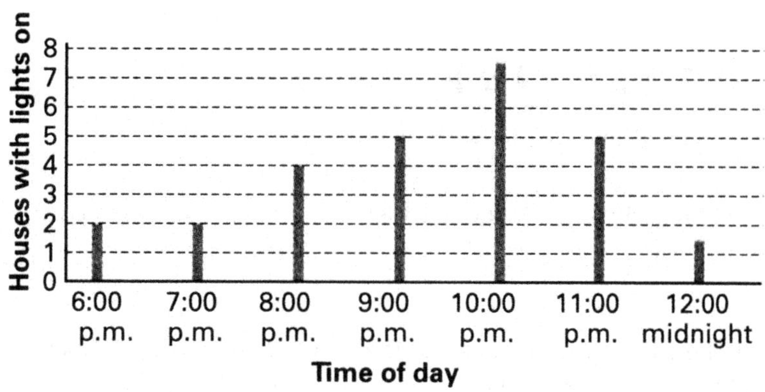

new word

Let children discuss with a partner what the **graph** represents. Choose some children to tell you what they think. Ask questions about the graph which focus on the changes over time, such as: 'How many more houses had lights on at 9 o'clock than they did at 6 o'clock?', 'Why do you think more houses had lights on at 10 p.m. than at other times?', 'Why might one house have had its lights on at midnight?'. Put the children in groups and ask them to make up a story about the graph on the board. That is, a story which explains why that number of houses had their lights on at those times. When all groups have finished give them time to share their stories with everyone else. As they are telling their story they should relate it to the graph.

Rainfall graph

Discuss what the graph is about. Talk about key aspects of the graph. Is the title accurate? How have the horizontal and vertical axes been labelled? What does mm mean? Ask questions which will help children to interpret the graph such as, 'Which is the wettest month?', 'Which is the driest month?', 'How much more rainfall is there in January than there is in July?', 'When is the wet season?', etc. In pairs children should make up some questions of their own about the graph and write them on paper. One child from each pair should ask one of their questions for the rest of the class to answer.

4B, p. 157

Graph display

Have the children copy the 'House lights' graph onto paper and write their stories about it underneath. Make a display in the classroom. Draw some other graphs on the board about changes over time and let children make up suitable stories about them. Have the children write their stories in both English and their vernacular underneath each graph.

OBJECTIVE 32.2

Sports days

Draw a chart like the following on the board:

Saturday	Sunday	Monday	Tuesday	Wednesday	Thursday	Friday

Ask the children to think about the sports they play and choose some children to tell the others about them. Are there some sports that children only play on one day of the week, or only on the weekend? Is there a sport which some children play every day?

As you say the name of each day of the week ask children to put up their hand if they usually play sport on that day. Count the number of hands and place that number under the corresponding week day.

Put the children in groups and ask them to discuss how they would draw a graph of how many children play sport on each day. Mention, or list on the board, key components you expect to see in their graphs such as a title and labels on the horizontal and vertical axes. Let each group tell the others how they would do it. Choose one of these ways and together draw a graph on the board to show the information. When you have made the graph ask some questions about it such as, 'How many more children play sport on Saturday than on Friday?', 'Which day of the week do most children play sport?', 'Is there a day when no-one plays sport?', etc.

How hungry?

Ask some children to tell you what time of the day they are most hungry. Are there times during the day when they are not hungry at all? Choose a child to tell you, starting at the beginning of the day, how hungry they are over the course of a day. As the child is talking make a graph about it on the board, showing the main times of the day e.g. getting up, breakfast, morning time, lunch, afternoon, evening, dinner time, and bedtime. Talk about the different ways children could show on their graph how hungry they are. Give each child a piece of paper and ask them to draw their own graph about how hungry they are over a day. As children are working, walk around and ask them questions about their graphs. When most children have finished choose some children to show their graph and explain it. Display the finished graphs in the classroom.

What is the graph?

Look at the first graph on the page and discuss with a partner what it might be the graph of. Accept any answers that children can justify. Also discuss the other graphs on the page.

4B, p. 158

Class graphs

Draw class graphs of other things that occur over a period of time, e.g. temperature each day for a week, number of hours of mathematics each day for a week, etc.

OBJECTIVE 32.3

One box means 5 children

4B, p. 159

Tell the children to look at the graph at the top of the page. Give them several minutes to work out what information is presented on the graph and then choose several children to tell what the graph is about, how the information is presented and what we can find out from the graph. If the children have difficulty interpreting the fact that each box represents 5 children, then you must explain this to them and go through each of the columns, asking how many children are represented by the boxes in each column. Can they work out how many children were involved in the **survey**? Is this number the same for each day on the graph?

new word

Tell the children to write in their books three pieces of information that they found out from reading the graph. For example 'Every day, more children walk to school than ride.' or 'Thursday is the most popular day for riding to school.' etc. Choose children to read out their statements and write these on the board. Be prepared to ask more questions if there are obvious pieces of information the children have not noticed about the graph. Ask the children what difference it would make to the graph if each symbol represented only one person instead of five. Tell them to read through the statements on the board with their partner. Then they should write the statements in their books and, beside each one, whether the statement would still be true if each symbol represented only one person.

Pictographs

4B, p. 159

The children look at the second graph on the page. Discuss why the baskets are lined up in a grid. What would happen if they were not like this? Tell the children to talk with a partner about the information presented in the graph. The children should also think of a name for the graph. After several minutes, invite children to tell the class their name for the graph and discuss which is the best name. Now ask the children to say what information the graph presents and how they worked this out. Write these statements on the board, numbering them consecutively. When most of the children have finished this, choose some to read out their answers for each statement and discuss them. Finally, ask the children to say whether they think it is a good idea to use one symbol to represent a number of objects instead of using one symbol for each item in the survey.

Newspaper graphs

Make a collection of simple graphs from newspapers, books or previous classes and have these available for the children to read and interpret by themselves or in small groups. Let them report their findings to the rest of the class.

OBJECTIVE 32.4

Drawing graphs

Survey the class on a topic where there are only two possibilities, for example: the children who have travelled on an aeroplane. On the board write the number who have travelled by plane and the number who have not. Ask the children to discuss with a partner the best way to graph this information. After a minute or so, invite children to tell their ideas to the rest of the class. Discuss what sort of pictures or symbols might be used to represent each group. Discuss the idea of using one symbol to represent a number of students. This will definitely make the graph quicker to draw, but will it also make the graph easier to read? Let the children give their ideas. On the board, draw a graph about aeroplane flight, using the children's ideas and the information you collected. Check that the graph represents the data you initially collected.

Highway survey

4B, p. 160

Explain that some children at another school made a survey of vehicles on the Highlands Highway. The graph is a record of what their results looked like. Choose several children to explain what information is recorded in the table. List this information on the board. Can they say how many cars, trucks, buses and motorcycles were on the highway? Now tell the children that they are to discuss with a partner how to present this information as a graph, using one symbol to represent 5 vehicles. Let them decide on a symbol for each form of transport and what to do about the odd vehicles that do not make up a group of five. Give each pair of children a large sheet of paper and let them draw a graph to represent the information in the table. Remind them to include a title, a key, labels for the two axes and a scale on their graph. Let the children show their completed graphs to the rest of the class. Make sure each graph correctly represents the information from the table. Refer to the list you made on the board if you need to.

Doing a survey

Provide opportunities for the children to conduct similar surveys, e.g. the people who pass our doorway in one day, and to make a graph of the results. You can also discuss the different ways the same set of information can be graphed. For example, you could classify the people into male/female, students/teachers/visitors, in hourly blocks to find the busiest times of day etc.

OBJECTIVE 32.5

Collecting data

Choose a popular topic or one that you are studying in your classroom, e.g. Food, Sport, Families and Hobbies. Ask the children to work in pairs and write down some questions about the topic so that the answers could be presented as a graph. (What is your favourite food? Who does the cooking/shopping in your family? What do the children in your class eat for lunch? etc) Ask some children to read out one of their questions and write these on the board.

Tell the children to discuss with their partner which question they will investigate and decide what information they need to collect to make a graph about it. The children should then decide how they will collect their data and how they will record it. Discuss each groups plans with them, then tell them to collect their **data** and make it into a graph. Tell them to use one symbol to represent a number of items when they make their graph. Give each group the opportunity to present their graph to the class.

Presenting information in a graph

Look at the scene of the village. Ask children to make a graph about any of the information shown in the picture (people in each family, number of the different types of animals, plants in the garden etc.)

4B, p. 161

UNIT 33

MONEY 2

This topic builds on the use of money and calculation with money (see *Teacher's Resource Book* 3, Unit 7) and also the litre as a measure of capacity (see *Teacher's Resource Book* 3, Unit 8).

ENGLISH WORDS/EXPRESSIONS USED

adjoining, non-adjoining

MAIN IDEAS

- **notes are used to pay large amounts of money**
- **there are different ways to make a particular amount of money**
- **calculations can involve money**

OBJECTIVES

The children will:

33.1 use the PNG note denominations to make various money amounts
33.2 calculate problems which link time and money
33.3 calculate problems which link capacity and money

SPECIAL METHODS AND STRATEGIES

Use sample notes to make this unit as realistic as possible.

SAMPLE ACTIVITIES

OBJECTIVE 33.1

PNG notes

Show the children each of the PNG notes that are in use (or use the pictures on Pupil Book 4B page 162). Discuss what is on each note, e.g. each note has a kundu drum and a bird of paradise on it. Talk about the colour of each note. When children have a good idea of the notes hide them and describe one of the notes to the children. Let children guess which note you are talking about. As you hold up each note ask children to tell you how many 1 kina coins they would need to exchange for the note. When you have done this ask why people need notes when they could use coins instead. Children should realise that it is for convenience, because it would be difficult to carry around fifty 1 kina coins in your pocket.

one K2, K5, K10, K20, and K50 note

How to pay?

State the problem 'I need to buy a radio for K45. I have some money in my pocket. What notes can I use to pay for it?' Put the children in groups to work this out. When most groups are ready let each group tell what notes they would use to pay for it. Are there different ways of doing this? Will I sometimes get change? Give the children another similar problem to work out, and allow time for them to share their suggestions.

Shopping

After looking at the pictures of each note and at the K1 coin in Frame A, children work out two different ways of paying for each of the items shown in Frame B. They should write the answers in their own books. When they have finished choose different children to tell how they would pay so that all children see that there are a variety of ways of making various money amounts.

4B, p. 162

Play purchases

Cut out pictures with price tags from magazines and papers. Put them in a tin or box. Children can take a picture from the box, read how much the item costs, and use 'play' money notes to purchase the item. Use the same items to practise giving change.

old magazines, newspapers, play money

OBJECTIVE 33.2

Paying the builder

Explain to the children that many people get paid by the hour. Tell them that this often varies depending on the type of work. For example, builders will probably get paid more per hour than cleaners. Discuss any examples the children are aware of where someone they know gets paid

paper 'play' money and coin rubbings

by the hour. Then present them with the following problem, 'If a builder gets paid K6.50 per hour, how much would he get paid for two hours?' Choose a child to give you the answer and write the cost for one hour and two hours on the board like this:

1 hour = K6.50

2 hours = K13

Then put the children into small groups to work out how much the builder would get paid for three hours, four hours, five hours. . . up to eight hours. Let them use 'play' money if they want to help them work it out. While groups are working check to see that they are using suitable strategies and understand what they are doing. Tell the children that the builder works five days a week for eight hours a day so that for a day's work the builder would get paid K52. Ask children to work in the same groups to calculate how much the builder would get for one week, two weeks, three weeks . . . up to ten weeks. When most groups have finished choose some to explain how they worked out the answers.

Calculating costs

4B, p. 163

Look at Frame A. A cleaner works five days a week for eight hours each day and gets paid K1.25 an hour. Children should use this information to help them complete the chart. They should copy the chart into their own books and complete the gaps. When they have finished this they should look at the telephone rates in Frame B. Discuss how the rates vary depending on where you are calling, and talk about what **adjoining** and **non-adjoining** mean. Children should now use the cost given per minute to work out the price of calls taking two, five, ten, twenty and thirty minutes. Children should copy the chart into their own books. Check as they are working. When most children have finished choose some to share their answers and methods of working out.

new word

Costs of phone calls

4B, p. 163

Use the information in the table in Frame B to make up some more questions about telephone charges relevant to your area. Include questions about calls from adjoining and non-adjoining zones.

OBJECTIVE 33.3

Cost of drinks

1 L drink bottle

Hold up a litre bottle of drink and ask if anyone knows how much it costs. Let some children have a guess before telling them the price. Write its price on the board next to the words 1 litre. Now ask children to work out how much two litres would cost. When they have done this choose some children to share the way they worked it out. Put children in groups to work out the cost of 3 litres, 4 litres, 5 litres . . . up to 10 litres of drink. They should write the costs in their own books. When most groups have finished allow time for them to share their answers and methods of working out.

Money problem solving

4B, p. 164

Children look at the prices of petrol, diesel, and engine oil, and use them to work out the answers to the questions on the page. Allow children to work in pairs to do this if they want. As they are working check that they understand what they are doing and give help where necessary. When most children have finished choose different children to give their answer for each question and show or tell how they worked it out.

Fuel for cars

Explain to the children that different cars hold different amounts of petrol or diesel. Generally, the bigger the car the more fuel it holds. Write on the board the following chart.

CAPACITY	COST OF PETROL	COST OF DIESEL
20 litres		
40 litres		
60 litres		
80 litres		

Put children in groups to calculate how much it would cost at current prices to fill the tank of cars that hold the above amounts of fuel or diesel. When most groups have finished share the answers and ways of working out, and write the answers on the chart.

Class party

Work out how many bottles of cordial you would need to buy for a class party and calculate the total cost.

UNIT 34

PAPUA NEW GUINEA GARDENS

This unit buids on the earlier work in this book. It draws on the skills and concepts covered in seperate topics throughout the year. It provides sample ideas for projects or investigations to allow the children to practise these skills and concepts in an integrated environment. Many of the activities in this unit are to be done outside the classroom.

MAIN IDEAS

- **that a range of skills and concepts can be applied to problems related to gardens**

OBJECTIVES

- **apply skills to real life situations**
- **work together to solve problems**
- **have opportunities to talk about Maths in English**
- **record their findings in a variety of ways**
- **be actively involved**

SPECIAL METHODS AND STRATEGIES

To prepare for this topic, you should:

- **read through the entire unit before you choose the activities for your class. You should include at least one of the longer activities.**
- **choose a garden in your community that will be suitable for the activities from this unit. Ask permission from the owner to regularly visit the garden with your class throughout this unit.**
- **make a large scale map of the chosen garden. You will need a large sheet of paper for this.**
- **have a selection of measuring equipment such as metre sticks, strings available for the children.**
- **have a metre square for each group of children. (Make the squares from sticks tied at the corners or from joined pieces of newspaper.)**

Many activities may take longer than a standard Maths lesson. Rearrange your timetable to provide blocks of time for teaching several maths lessons together. An estimate of the time required for these activities is included.

SAMPLE ACTIVITIES

Garden map

This activity should take 3-6 lessons depending on the size and location of the garden. Take the children to the chosen garden. Let them look at the main features of the garden. Tell them to take note of the overall shape of the garden. Tell the children to sketch the overall shape of the garden. In groups have them measure the lengths of the edges of the garden and mark these on their sketches.

Garden, measuring tools, pencils, paper, pad or card.

Tell the children that they are going to map the garden. They are to choose some of the main features, find where they are located in the garden and then mark them on their sketches. Have the children measure the distance of the main features from two different edges of the garden so that they can place them accurately on their maps. The children should prepare a table like the one below and complete it.

FEATURE	DISTANCE FROM EDGE 1	DISTANCE FROM EDGE 2	SHAPE OF FEATURE	SIZE OF FEATURE
old tree stump	16 paces	4 paces		2 paces inside
corn	7 paces	2 paces		3 x 2 paces

The children will have to measure the size of each feature and sketch it. Back in the classroom each group reports on their results. The class combines the information they gathered and from this information each group draws a scale map of the garden. They should include a key and a scale.

If possible groups return to the garden with their maps and check them against the real thing. They can then make changes to their maps if necessary.

Describe a garden

In groups, children look at the different types of gardens shown in the photographs. Discuss the overall shape of each garden. Ask children to talk about how each garden compares with the one the class visited?

4B, p. 165

Parts of gardens

This activity could take 1-2 lessons. Take the children to the chosen garden. Ask them to choose the food crop that occupies the largest area of garden land, and the one which uses the smallest area of garden land. Ask the children to suggest ways of checking their answers. Let them try out their strategies. Report back to the class. Let each group first estimate and then measure the area of a different food crop. Report back to the class. This information can be added to their garden map if they have drawn one.

garden, measuring tools, square metre, pencils, paper, pad or card

Ordering by size

Look at the map. Ask the children to estimate which of the plots in this garden is the biggest, second biggest and so on down to the smallest. They should list the plots in their books in order from largest to smallest. When they have done this, tell them to use the base 10 units or similar to measure each of the plots and to write its area in their books. Did they arrange the plots in the correct order? Check their answers.

4B, p. 166

Counting seedlings

counters

Briefly discuss the garden plots that are shown. Look at the first garden plot, and ask the children to tell you how many seedlings are planted in it. When all children have worked this out ask a child to tell you the answer and write it on the board (20). Now ask some children to tell you how they worked this out. There are various ways to do this such as: counting by one's, counting by two's, counting one row and doubling it, multiplying 2 by 10, multiplying 10 by 2, etc. Accept all methods that lead to the correct answer. Have the children work in pairs to write two multiplication number sentences for each of the other garden plots shown on the pupil's page. When they have finished choose some pairs of children to give their answers.

4B, p. 167

Garden number sentences

Visit the garden you chose in *Garden map* and write some multiplication and division number sentences about the plants in the plots.

Distance between plants

garden, measuring tools, pencils, paper, pad or card

This activity may take 2-4 lessons. Take the children to the chosen garden. Let each group choose some plants that are planted in one area, e.g. corn, kaukau. Ask, 'Are all the plants spaced the same distance apart? Are the plants in rows? Are the plants in a grid?' Each group should make a map of their part of the garden.

Number some (or all) of the plants in this part of the garden. For each plant measure the distance to the next plant. Use these measurements to complete a table like the one below.

PLANT	DISTANCE TO NEXT PLANT
1	20cm
2	26cm
3	15cm
4	

Can the children identify any patterns in the measurements?

Each group should list: the longest measurement, the shortest measurement, the most common measurement. What is the difference between the longest and the shortest measurement?

Ask each group to discuss what measurement or measurements they could use to represent the plant spacing for their plants, e.g. range (longest – shortest) and the mode (most common) mid-point of range or some other suitable middle value. Each group reports to the class. Ask, 'Why are different plants spaced differently?'

What plants might their be?

Look at the rows in each plot. Discuss what scale has been used to show the distances between plants in each plot. Discuss what plants might be planted in each plot, based on the information children have found about planting distances.

4B.p. 168

Planting a garden

Tell the children to ask their parents, or the people who do the planting in their family garden, how they make equal distances between plants as they are planting. Do they use a body part as a 'measurer' (e.g. a forearm, three foot-lengths), do they step out a number of paces for larger plants, or do they make a 'measurer' such as a knotted length of string or a stick with notches in it. Have the children report back to the class or invite an adult to come and tell the children about planting in the garden and the distance apart needed for each kind of vegetable or fruit.

Lines and scales

Have children mark a scale on their garden maps. Look at the number line at the top of the page. Point out that 1 cm on this number line equals 20 cm in the bean plot, and show this in the classroom by measuring out 20 cm distances and using counters to represent the bean plants. Discuss why no plant is placed at the start or end of the row. Now let children in pairs use rulers to work out how many beans we could plant in the row. When most pairs have an answer make time for some of them to explain how they worked it out. Look at the other number lines on the pupil page. Children use the information shown to work out what scale is used on each, and then use the scale and a ruler to work out how many plants will be in each row. Look at the first number line which shows a planting distance of 30 cm apart. Discuss the types of plants that this number line might be referring to. Do the same for the other number lines. Use the number lines on the pupil's page to work out how long each of the planting rows would be.

4B, p. 169

Garden calendar

This activity may take 2–3 lessons. Put the children in groups of four and give them their sheet of paper with a 40 cm diameter circle drawn on it with the centre point marked. Tell them that they are going to make a

large paper, coloured pencils, crayons, pastels, paste

calendar for a garden. Their first job is to mark the months on their calendar. Show them how to use their rulers to divide the circle into quarters by ruling a vertical line through the centre and then a horizontal line at right angles to this.

Ask the children how many sections they will have to make in each of the quarters to have 12 months in all. (3) Let them mark the circumference of each quarter into thirds and then join these marks to the centre point

Tell them to write the name of each month just inside the circumference of each of the 12 sections on their calendar. Now ask the whole class to think of things that happen throughout the year that affect the garden. Make a list of these on the board. They will include things like the wet and dry seasons and in which months these occur, the cycle of winds, when various crops are planted and harvested, when the weeding and different maintenance jobs are done, and any festivals that are held to mark special days associated with gardens or produce. If there are special market days you could list these too.

Now tell the children to work out in their group how this information can be shown on their calendar and to decide how to share the workload. Provide them with coloured pencils, crayons, pastels, paper and paste, etc. Allow time for them to complete their calendar. Give help where needed. When the calendars are complete, let each group present theirs to the whole grade and explain how they represented the information. Compare the calendars by discussing effective ways of presenting the information.

The Garden Game

Give each group a dice and four different counters. Ask a child to read the instructions at the bottom of the page and then let them play the game in their groups.

dice, counters

4B, p. 170

A plant growth investigation

This activity will take a short time every second day. Put the children in groups of 3 or 4 and give each group 5 markers. Tell them to label their markers so they know they belong to their group. Now take the children to the garden and ask them to choose five different kinds of plants. They should list the names of their plants, e.g. corn, tomato, in their notebooks. Try to choose young plants which will

markers, rulers marked in centimetres

show some growth over a short period. They should place a marker beside each of their five plants. Each group should measure the height of each of their plants and record the height of the plant and the date the measurement was taken in their notebook beside the name of the plant.

Back in the classroom decide how long you are going to continue to take measurements, e.g. 3–5 weeks. Help the children to rule up 5 sheets of paper to make graphs of the growth rate of their five plants. Discuss what scale they will use for each plant as this may vary for different plants. For example corn grows more quickly than pineapple. When they have drawn up their graphs, they should mark in today's measurements. Then store their graphs somewhere so that they can be added to every second day.

Have the children return to the garden every second day, measure their plants and record the measurements and date in their notebook. Every second day they should record the new measurements on their graphs. At the end of the 3–5 week period you will need to hold a class discussion where the children show the other groups their completed graphs and note similarities and differences between plants.

Graph of plant height

Give out the rulers. Ask what the graph is about. Ask questions like, 'How high does the pawpaw tree grow?' Make sure the children can use their ruler and the scale to find the answer. Let the children write the answers to the questions in their books.

4B, p. 171

Working in the garden

a demonstration clock

This activity should take a few minutes each day for a week. The table will work best if it is done at one of the busy times of the year like planting or harvesting time. Talk with the children about who works in the garden. Tell them to make a list of the people who worked in their garden last week. They should write the list down the side of a page in their books. Show them how to draw up a table like this for entering the number of hours each person worked in the garden each day.

	Mon	Tues	Wed	Thurs	Fri	Sat	Sun	TOTAL
Mother								
Father								

Each day during the week have the children ask the relatives listed in their table, how long they spent in the garden that day. Children record the answer in their table. Ask them to add up the total number of hours each person spends in the garden. Then ask them to add up the total number of hours of work done in their garden altogether by all the people who help.

Ask the children to swap books with a partner. The children should read the table in their partner's book, to see who does the most work, who works most often, who doesn't help at all etc. Ask the children to take turns to tell their partner four things they found out about who works in their partner's garden. Choose some children to report their observations to the whole class.

A 30 hour week

4B, p. 172

Tell the children to look at the table and work out what information it tells them. (Who is in the family, who is likely to do the most and the least work, how many hours the mother works etc.) They copy the table into their books and write in what hours the other family members should work to make the total hours worked in the garden add up to 30.

Changes in garden workload

Discuss with the children how the workload in the garden changes throughout the year. Let them refer to the calendar they made in '*Garden Calendar*' and discuss which are the busiest and the quietest times of the year. Based on the total hours worked in their garden of the past week, could they estimate how many hours work would be done in the busiest month and in the quietest month of the year?

Estimating height and width of trees

a notebook and pencil , 2 m bamboo poles or sticks

Put the children in groups of 3 or 4 and then take them to a tree in the school grounds. (A tree that is about 4 metres tall would be ideal.) In groups they estimate (or guess) how high the tree is and write this down. When all have an estimate ask each group to say what it is and to explain what helped them to make their estimate. For example, they may have used their own height as a guide, or a building which they know the height of, etc. After all groups have described their method allow them to alter their estimate if they wish. Give each group a 2 m bamboo pole. Then, using the 2 metre pole, ask a child from each group to stand near the tree and hold the pole so that one end is resting on the ground. Have the other children in the group stand back where they can see how many bamboo poles would be needed to reach the top of the tree. In this way each group works out how tall the tree is. Then they compare this estimate with their earlier ones. How close were they?

Choose three or four more trees close to the school and ask groups to use the bamboo poles to estimate how tall these trees are. Give groups time to do this. When they have finished compare the findings of the different groups.

Return to the first tree and ask children to discuss how they could estimate the width of the widest part of the tree. Give groups time to discuss this and then ask each group to tell their ideas to the others. Decide which of the methods would be most suitable and give groups time to use this method to estimate the width. (A method which involves marking lines on the ground below the widest part and then using a bamboo pole or their pace to get an accurate estimate would be most suitable.) Let groups use this method to estimate the width of some other nearby trees and write their estimates in their notebook. Compare the estimates of each group checking where there are any big differences.

Judging heights

4B, p. 173

Discuss situations where it is useful to know the height and width of trees. For example, to choose a young tree to plant in a certain space; when planting trees in rows we need to know how much space to leave between them; the width of trees gives us an indication of where the roots spread under the ground. Children should look at the pictures on the page and use the length shown on each pole to estimate the height and width of the tree next to it. When estimating the width it would be useful for children to use a strip of paper to mark the pole length against so they can then hold it horizontally against the tree's leaves. When most children have finished check their work as a class.

A mathematical tree

a portable board or large piece of paper, measuring instruments to measure metres and centimetres, pan balances and weights

This activity will take 2–5 lessons. Take the children to a nearby tree. Ask the children to look at the tree and to think about all the mathematical things they could find out about the tree, for example, how high it is or how many leaves it has. As children make suggestions write them on the board or paper. (Other possible suggestions include: size of leaves, weight of leaves, distance around the trunk, diameter of tree, amount of space tree needs, how much shade tree provides, number of branches, distance of lowest branches from the ground.) When the children have run out of ideas return to the classroom.

Put the children in groups of 3 or 4. Ask each group to choose something on the list that they would like to investigate. It does not matter if more than one group chooses the same thing but try to make sure that everything on the list is chosen. Some things will not take very long to investigate so groups who choose these should choose something else to investigate as well. When all groups know what they are investigating, give them time to plan how they are going to do it. This planning is a very important step and groups should tell you how they are going to do their task before they start.

When you are satisfied that each group has a suitable plan assist them to find any materials they need and take them outside to an area where they can carry out their investigation. Allow time for them to do their investigation, helping where necessary. When groups have finished they can explain to the rest of the class how they carried out their investigation and tell what they found out.

How many leaves?

After looking at the picture of the tree they should work in pairs and discuss how they could estimate how many leaves it has. Give time for pairs to try out their ideas. Make sure children understand that they have to estimate - not count each leaf. When children have made an estimate allow time for them to explain to the others how they worked it out. Decide which method was most suitable.

4B, p. 174

Weighing garden produce

fruit and vegetables - 1 kilogram of each, a balance and $\frac{1}{4}$ kg, $\frac{1}{2}$ kg and 1 kg weights

Show the children the fruit and vegetables. If possible, pass them around so that the children can feel the weight of each item. When they have had a chance to handle the fruit and vegetables, ask them to say which they think is the heaviest. Accept 2 or 3 guesses and put the items named to one side. Ask the children to use the balance to find out which of these is the heaviest. Choose two children to demonstrate their method. Repeat this procedure to find which is the lightest item in the collection. Now ask the children to draw up a table like this in their books.

Less than 1 $\frac{1}{4}$ kg	More than $\frac{1}{4}$ kg and less than $\frac{1}{2}$ kg	More than $\frac{1}{2}$ kg

Ask them to draw, or write the names of three or four items that belong in each category. When most children have finished doing this, give out the balances and weights and let the children work in groups of 4 to check their answers. They will need to take items from the collection to weigh and return them when they have finished. When they have finished checking, choose children to say what items from the collection belong in each of the categories. (If you only have one balance, do this step with the whole class, letting children take turns to weigh the different items.)

Now ask the children to draw five items from the collection in their books and to write beside each one how many would be needed to make up 1 kg. When most of the children have finished doing this, ask a child to tell the class one of their predictions, e.g. 'I think it would take 4 avocados to make 1 kg.'. Choose children to use one of the balances and a 1 kg weight to check whether the prediction is correct. Continue in this way until all their predictions have been checked.

Weight of fruit

Ask a child to explain what the balances show about the weight of each piece of fruit. (For example, if 4 mangoes weigh 1 kg, how much does each mango weigh?) Ask children to read through the questions and make sure everyone knows what they have to work out. Give children time to write the answers to the questions in their books. When they have finished, choose some children to read out their answers so that the others can check.

4B, p. 175

At the market

This activity should take 1–3 lessons. Take children to the market. Buy some of the market produce to bring back to the classroom. Try to bring back 5 to 10 different fruits or vegetables. The children should use the balances to find the weight of the quantities sold. Let the children in groups fill a table like the one below.Children may need help to fill in some of the spaces.

market, market produce, balances

FRUIT OR VEGETABLES	SOLD AS	PRICE	HOW MANY?	WEIGHT/SIZE	COST OF 2 kg	COST OF ONE
tomatoes	pile	K2	8	1 kg	K4	25t
mango	one	50t	1	500g	K1	50t
bananas	pile	K1	3	400 g	K5	33t

At the market 2

If it is not possible to actually visit the market, or bring real fruit and vegetables into the classroom, use the information on the pupil page to complete a table like the one in *At the market*.

4B, p. 176

A day in the garden

Ask children to recall a day when they worked in their garden with their parents. Using either the vernacular or English, discuss what time they got up, what they had to do before they left for the garden and what time they set out for their garden. Continue like this until children have outlined all the events of the day. As you go, ask children to show the times that they mention on the demonstration clock and to say whether the time is a.m. or p.m.

a demonstration clock

A day in the garden 2

4B, p. 177

Tell the children to look at the pictures on the page. Choose children to describe what is happening in each of the pictures. Choose one child to tell the story of the mother's day. Ask the children to draw in their books a clock-face to show what time the events in each picture might be happening. They should label each clock-face with the time it shows and whether it is a.m. or p.m. Check while the children are working to make sure they are drawing the clock-faces correctly and showing the correct position for the hands. Choose children to say what time they thought each picture was happening. Now ask the children to draw six pictures to show what they would be doing at the same time as each of the six pictures on the Pupil page. They should number each of their pictures to correspond to the pictures on the Pupil page. When they have finished, let some children share their pictures with the rest of the class.

Sorting fruit and vegetables

fruit and vegetables, three circles made of

Place a circle of string, and the fruit and vegetables (one of each type) on the floor. Choose a piece of fruit or a vegetable, e.g. a banana and place it in the string circle. Choose another piece of fruit that has to be peeled and place it with the banana. Ask children if they know why you have put those fruits in the circle. Continue to select fruits or vegetables which have to be peeled until someone tells the 'rule'.

Using a different 'rule' select other fruits and vegetables and put them in the string circle. Other children observe what you are doing and tell you the rule used. For example, fruit or vegetables with a stone; those that can be cooked before they are eaten; those that grow on vines; etc.

Ask some children to choose a 'rule' of their own and do the same activity as above.

Place two of the string circles on the floor. Have the children as a class suggest how they could divide the collection of fruit and vegetables into two groups and then do this by placing them in the string circles. They should then describe the groups. Can they do this another way? (For example, if they first sorted them into fruits and vegetables they might then sort them into smooth or rough skin groups.) Now place the third string circle on the floor and have them sort the fruit and vegetables into three groups and describe the groups.

Now invite a child to sort the collection into groups and invite the others to explain the 'rule' that was used. Let other children have a turn. Each child could draw in their books a way of sorting the collection.

Fruit groups

4B, p. 178

Look at the first diagram. Children have to work out what the 'rule' is for the fruit and vegetables in the circle, e.g. food that grows under the ground. They should write down the names of other fruits or vegetables in Frame A that could be included in the circle. Do the same for Frames B and C, e.g. foods that are yellow, and foods that have a hard casing.

Teaching Materials

On the following pages of this book there is a description of the materials you will need to make to be able to teach some of the lessons in the Grade 4 Mathematics Program. The following is a set of instructions to help you make these materials.

HEXAGONAL SPINNER

A hexagonal spinner can be used instead of dice. You will need: firm cardboard, scissors or sharp knife, rulers, 5 cm stick, glue

1 Using a ruler, accurately mark out a regular hexagon on firm cardboard. (All sides must be the same length.)
2 Draw in all the diagonals.
3 Write in the numbers 1 to 6.
4 Use a small (5 cm) sharp stick to make a handle and spinning point.
5 Glue this stick in the centre of the hexagon.
6 Hold the handle between your thumb and another finger and twist to set the spinner spinning. The number the spinner rests on when it stops spinning is the number thrown.

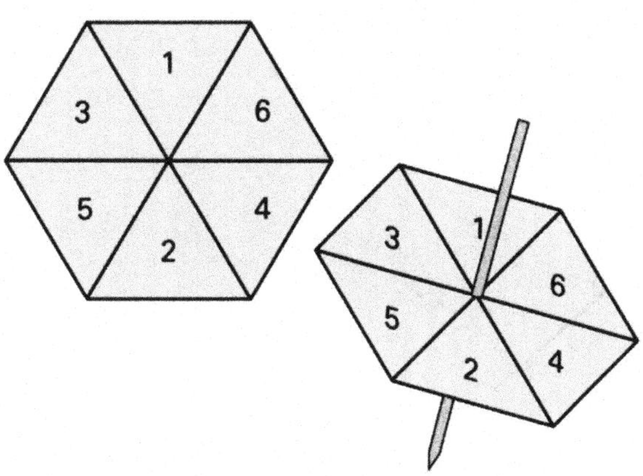

SMALL NUMBER CARDS

You will need firm cardboard.

1 Cut the cardboard into 4 cm × 4 cm squares. Make as many squares as you can (at least 150).
2 Number the cards so that you have at least ten sets of cards numbered 0 to 9. Leave the remaining cards blank so you can write other numbers on them as you need them.

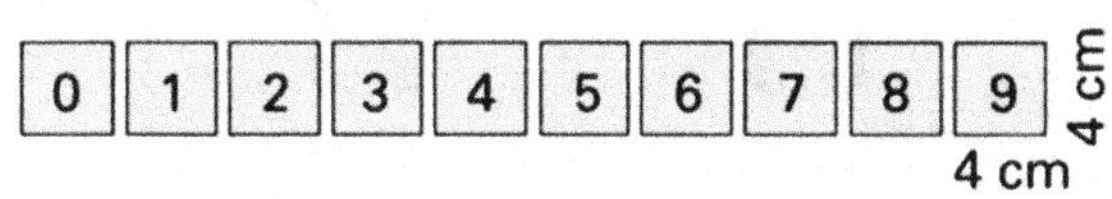

TIMES TABLE CARDS

You will need paper or cardboard.

1 Cut the cardboard into rectangles, at least 3 cm × 8 cm. You will need eleven cards for each times table. If you make a set for each of the times tables (2 times table to 10 times table), you will need 110 cards altogether.
2 Write a times table fact on each card. Below is an example for the two times table cards.

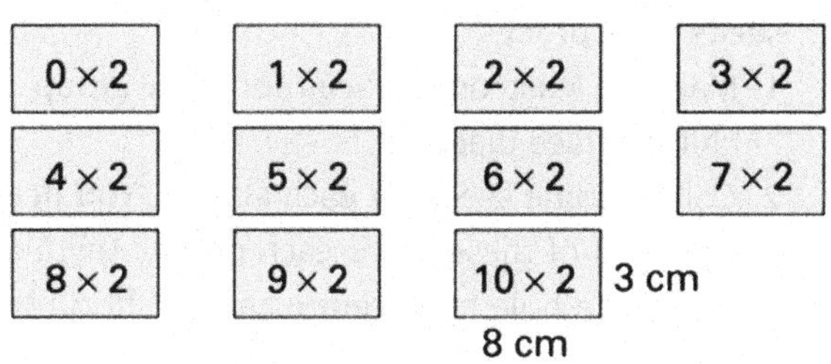

BALANCES

Here are some suggestions for making balances. Choose the one that you find easiest. You will need at least six balances for your classroom. It is even better if you can make more. You will need: string, bamboo, tin cans (plates or coconuts, halves can be used instead), coat-hangers (bamboo can be used instead)

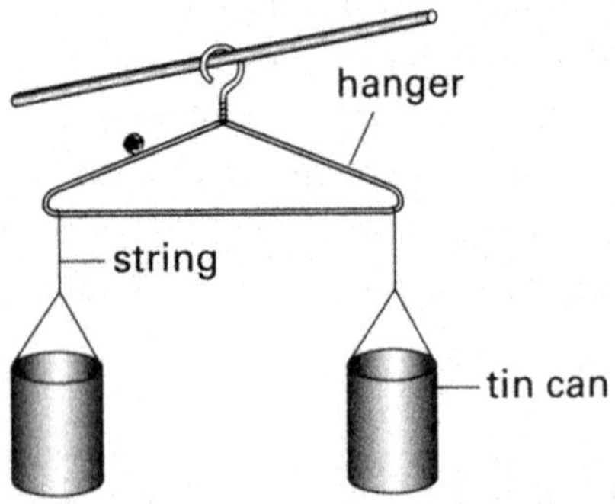

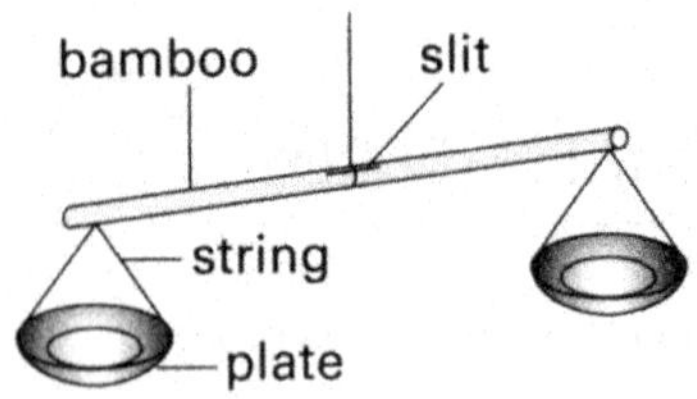

ABACUS (FOUR PRONGS)

You will need:
four sticks for the prongs, a block of wood, glue, seeds to fit prongs

1 Make at least one abacus for each group of children (see diagram).
2 Collect some seeds for each abacus. You need to fit ten of these on to each prong. Anything that has a hole in its centre and will fit on the prong is suitable, e.g. beads, small bamboo pieces, seeds, shells.
3 Glue the sticks to the block of wood or drill holes in the block of wood to hold the sticks. The prongs should fit firmly into the holes.
4 You can also use sticks pushed into the damp soil.

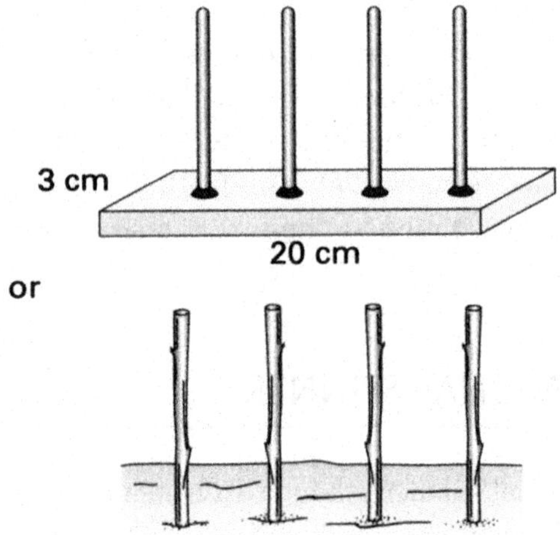

COIN RUBBINGS

You will need:
one of each type of coin, paper and lead pencils

1 Place a piece of paper over one side of the coin.
2 Rub a pencil across the surface until the coin markings show on the paper.
3 Cut around each rubbing.
4 To make the coin rubbings stronger, glue them on to cardboard of the same size.

Make as many of each coin as you need.

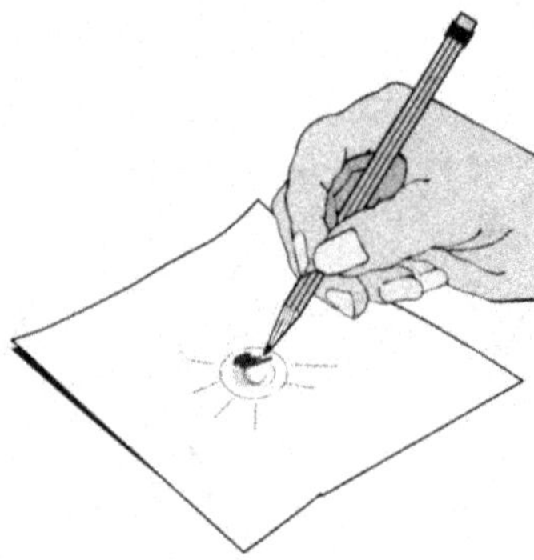

PLACE VALUE SQUARES

You will need cardboard of firm paper.

1 Cut the cardboard into squares:
 4 cm × 4 cm, 5 cm × 5 cm, 6 cm × 6 cm, 7 cm × 7 cm.
2 Write the number 1 on the 4 cm × 4 cm cards.
3 Write the number 10 on the 5 cm × 5 cm cards.
4 Write the number 100 on the 6 cm × 6 cm cards.
5 Write the number 1000 on the 7 cm × 7 cm cards.
6 Make at least ten of each card for each group of children.

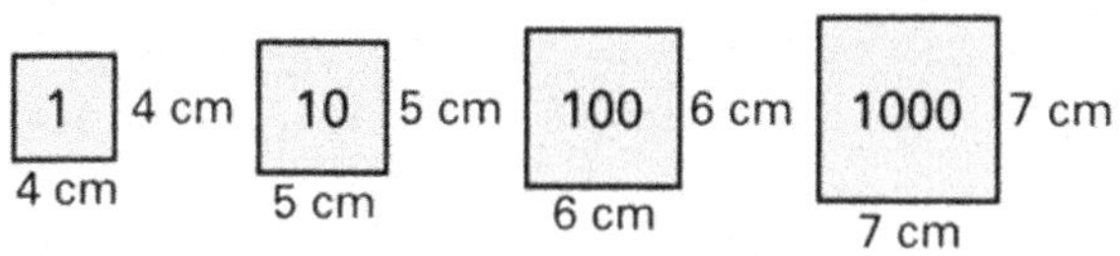

SPEAR CARDS

You will need paper or cardboard.

1 Make a template for each set of the four spear card shapes as shown in the diagram.
2 Trace the outline of each set of templates on to the cardboard or firm paper to make nine of each shape. Cut them out.
3 Write the numbers from 1 to 9 on the smallest cards, 10 to 90 on the next set of cards, 100 to 900 on the next set of cards, and 1000 to 9000 on the largest cards.
4 Make sets of each of the four spear card shapes for each group of children.

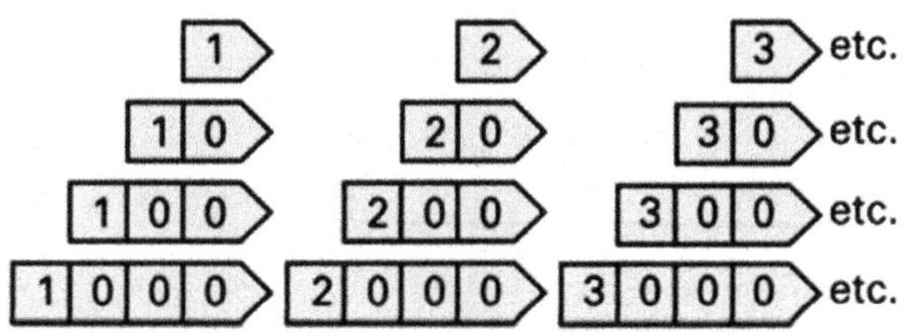

DEMONSTRATION CLOCK

You will need:

cardboard for the clock face (at least 20 cm in diameter), and for the two hands a nail or pin to join the hands to the clock

1 Cut out a large circle from the cardboard. Make sure its diameter is at least 20 cm. This will be the clock face.
2 Write the numbers from 1 to 12 in their correct positions on the clock face, as shown in the diagram.
3 Draw the minute marks on the clock face.
4 Make a long hand (the minute hand), about 10 cm long.
5 Make a short hand (the hour hand), about 7 cm long.
6 Find the centre point of the circle and join the hands to the clock face by pushing the nail or pin through the short hand, the long hand, and then the clock face. Bend the end of the pin over at the back of the clock so that the hands are able to be moved only when pushed around the clock.

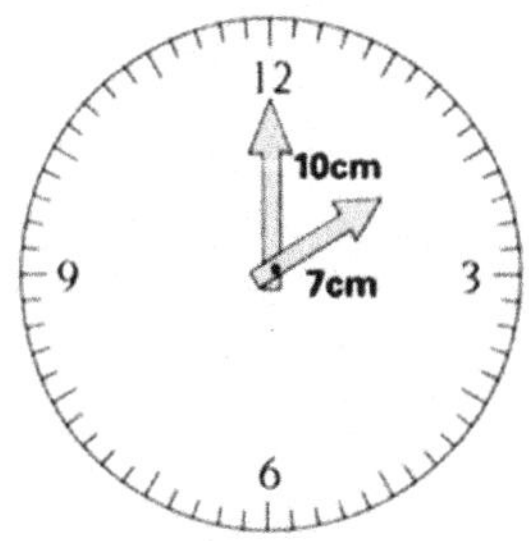

TANGRAMS

You will need cardboard or paper.

For the 5-piece tangram:

1 Cut a large square of cardboard—at least 10 cm × 10 cm.
2 Draw fine vertical and horizontal lines showing the centre of the square.

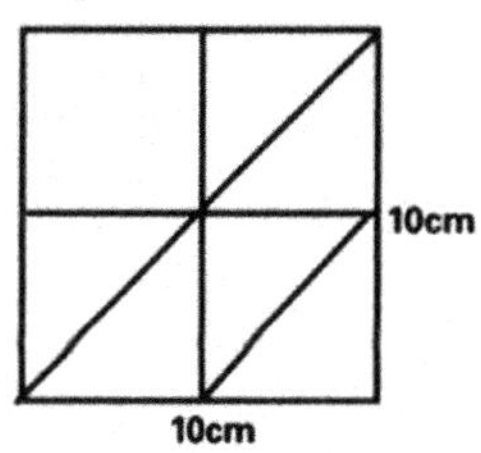

3 Rule and then cut along the heavy lines as shown in the diagram.

ANGLE TESTER

You will need:
two thin strips of card about 15 cm long
a paper fastener (or a drawing pin, nail, etc with plasticine or cork to put over the point)

1 Cut two thin strips of card as shown in the diagram.
2 Make a hole near the end of each strip.
3 Join them together by pushing a paper fastener through the holes. If you do not have a paper fastener push a drawing pin or nail through the holes and place a piece of plasticine or cork over the sharp point.

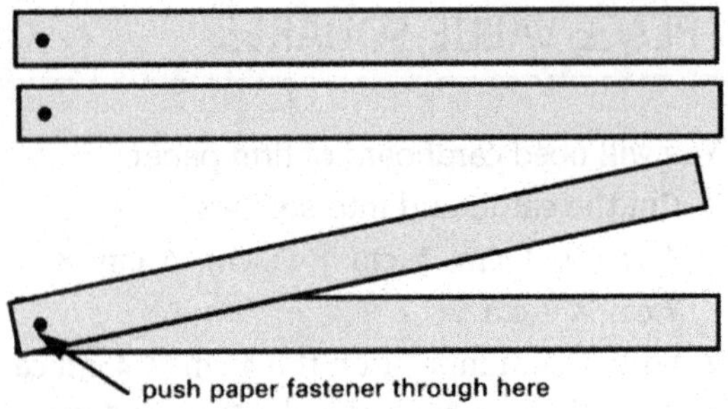

NETS

In Unit 16 you need a triangular prism and a hexagonal prism. The following diagrams show you how to make these. Copy each one on to light card, cut around the edges, fold on the dotted lines, and paste the flaps to join the shape together.

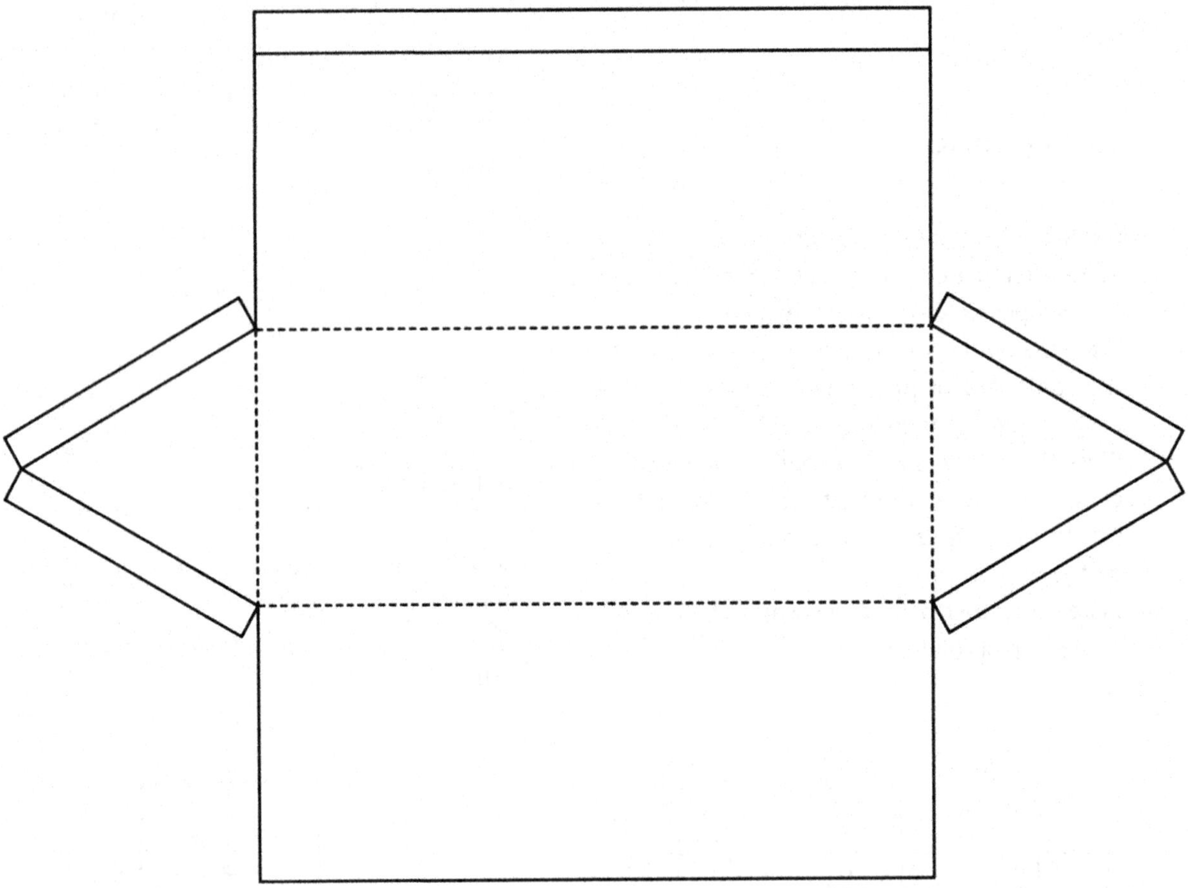

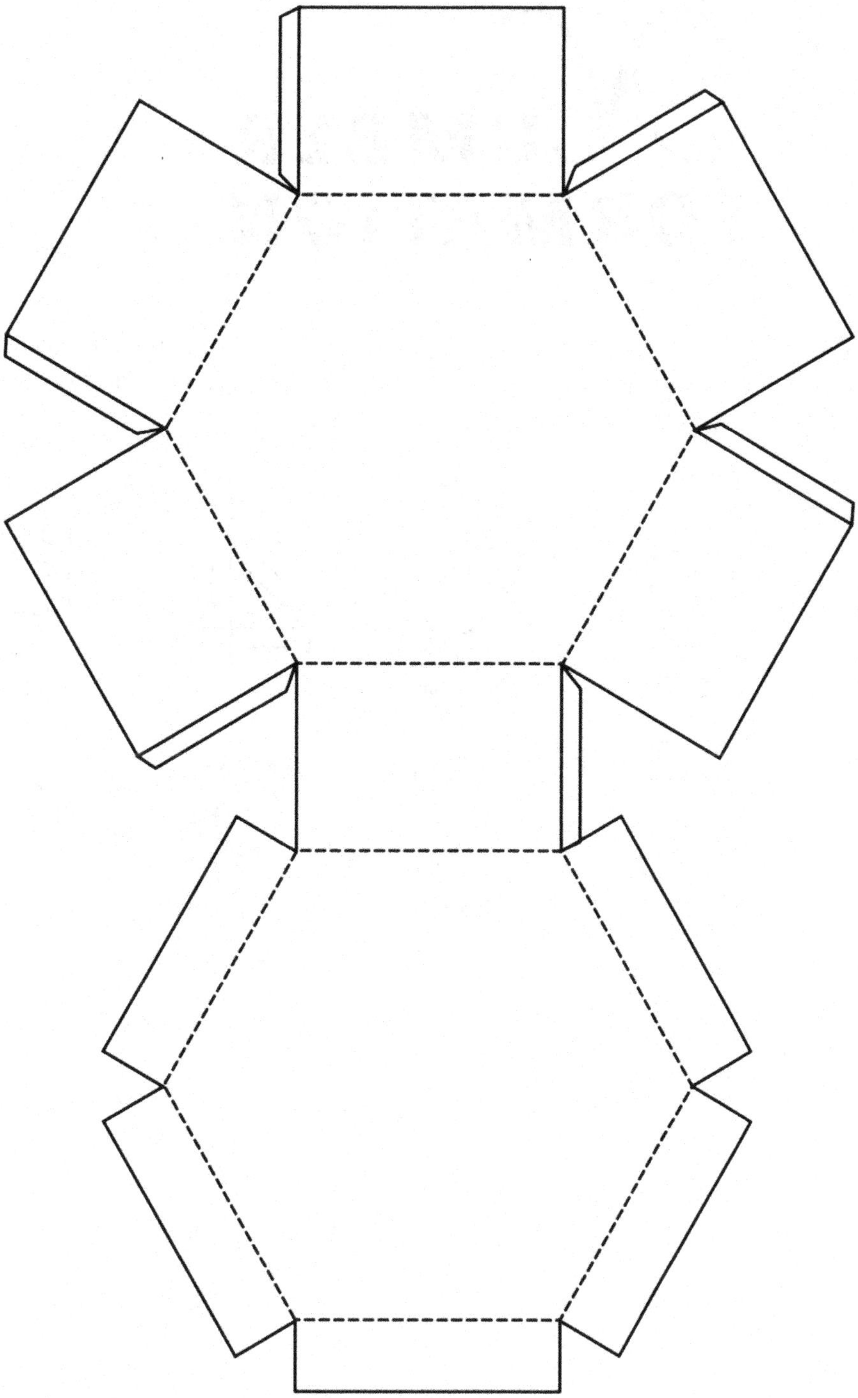

NUMBER FORMATION

When you write numbers on the blackboard, or anywhere else, make sure you write them properly, always starting in the correct place and taking care with their formation. The following is the recommended way of writing numbers.

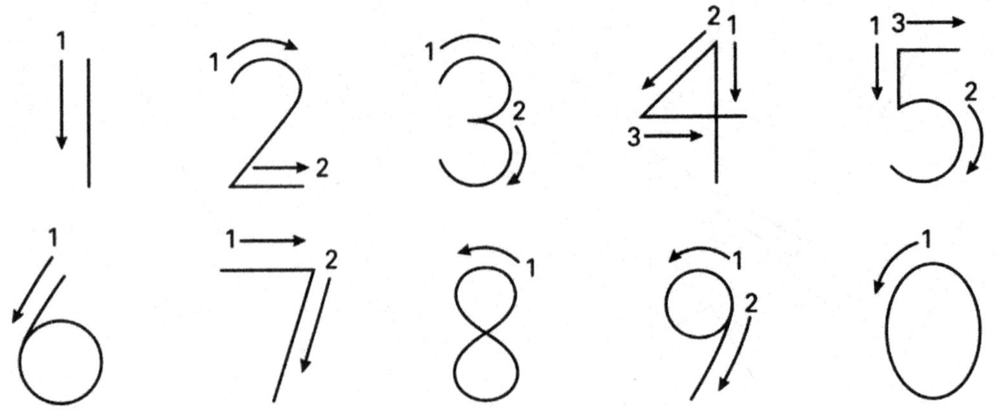